H. Duddeck, W. Dietrich

Structure Elucidation by Modern NMR

A Workbook

With a Preface by
J. B. Stothers

Steinkopff Verlag Darmstadt
Springer-Verlag New York

Prof. Dr. H. Duddeck
Dr. W. Dietrich
Fakultät für Chemie
Ruhr-Universität Bochum
Postfach 102148
4630 Bochum, FRG

CIP-Titelaufnahme der Deutschen Bibliothek

Duddeck, Helmut:
Structure elucidation by modern NMR : a workbook / H. Duddeck ; W. Dietrich. With a pref. by J. B. Stothers. –
Darmstadt : Steinkopff ; New York : Springer, 1989
    Dt. Ausg. u. d. T.: Duddeck, Helmut: Strukturaufklärung mit moderner NMR-Spektroskopie
    ISBN 3-7985-0787-2 (Steinkopff) brosch.
    ISBN 0-387-91348-3 (Springer) brosch.
NE: Dietrich, Wolfgang:

Copyright © 1989 by Dr. Dietrich Steinkopff Verlag GmbH & Co. KG, Darmstadt
Chemist Editor: Heidrun Sauer – Copy Editing: Marilyn Salmansohn – Production: Heinz J. Schäfer

Printed in Germany

Type-Setting: E. Pendl, Heidelberg – Printing and bookbinding: Weihert-Druck GmbH, Darmstadt

Foreword

For several years we have been organizing seminars and workshops on the application of modern one- and two-dimensional NMR methods at the faculty of chemistry in the Ruhr-University Bochum, FRG, and elsewhere, addressing researchers and graduate students who work in the field of organic and natural products chemistry.

In 1987, we wrote a workbook (Strukturaufklärung mit moderner NMR-Spektroskopie, Steinkopff, Darmstadt, FRG, 1988) in German language based on our experience in these courses. Many of the exercises described therein have been used in such courses and some of them have been shaped by the participants to a great extent. The response of readers and discussions with colleagues from many countries encouraged us to produce an English translation in order to make the book accessible to a wider audience. Moreover, the content has been increased from 20 exercise examples in the German, to 23 in the English version.

This book could not have been written in the present form without the help of a number of colleagues and, therefore, we acknowledge gratefully the generous supply of samples from and useful discussions with B. Abegaz (Addis Ababa, Ethiopia), U.H. Brinker (Bingham, New York, USA), E. Dagne (Addis Ababa, Ethiopia), M. Gonzalez-Sierra (Rosario, Argentina), J. Harangi (Debrecen, Hungary), S.A. Khalid (Khartoum, Sudan), A. Lévai (Debrecen, Hungary), M.A. McKervey (Cork, Ireland), M. Michalska (Lodz ,Poland), E.A. Ruveda (Rosario, Argentina), G. Snatzke (Bochum, FRG), L. Szilágyi (Debrecen, Hungary), G. Tóth (Budapest, Hungary), P. Welzel (Bochum, FRG), J. Wicha (Warsaw, Poland) und K. Wieghardt (Bochum, West Germany).

We also thank M. H. Kühne, Mrs. D. Rosenbaum, M. Gartmann and Mrs. E. Sauerbier for their committed cooperation, their assistance in the measurements, and preparation of figures.

Inspite of painstaking efforts mistakes can hardly be avoided. We are always grateful for any response from readers to correct or improve the text.

If we have been successful in conveying an impression of the wealth of information offered by modern NMR,then the book has satisfied its goal.

Bochum, FRG, February 1989   *Helmut Duddeck*
                                                               Wolfgang Dietrich

Preface

(translated from the original German edition)

The history of nuclear magnetic resonance (NMR) is characterized by a number of significant technical achievements. The latest progressive step is the invention of the two-dimensional NMR spectroscopy, which, with its concept of time evolution, has given rise to the development of numerous, also one-dimensional, techniques. It is fortunate that, at the same time, cryomagnetic technology has reached a high point in its development. Consequently, high field NMR spectrometers are now standard equipment for university chemistry departments and industrial laboratories, so that larger and more complex molecules can be investigated with respect to their structure, dynamics and reactivity.

It is not an exaggeration to say that applied high-resolution NMR spectroscopy has been revolutionized by the two-dimensional methodology. Previously, measurement of the NMR spectrum was confined to standard experiments involving spin excitation and signal registration; little allowance was made for variation. Now a number of experiments with different objectives and various levels of sophistication are available, often making it difficult to decide which of these experiments can reliably supply the desired information in as short a recording time as possible. This problem can only be solved by chemists who are well versed in the new techniques.

It is therefore fortuitous that Helmut Duddeck and Wolfgang Dietrich have used their experience gained in the NMR laboratory of a large chemistry department to fill the gap between spectroscopists and chemists working synthetically. In this volume they tell us about modern one- and two-dimensional NMR experiments on molecular structure and pertinent NMR analysis, and in so doing, arouse interest in such experiments in general. Moreover, they elucidate the potential and the limits of these new techniques. Thus, they have created a workbook that concentrates on the essential methods already approved in practice. The book follows the pragmatic tradition of the American textbook, which regards the "Aha! experience" gained by working with practical examples as being as important as the study of consistent, theory-based treatments. The book contains excellent illustrations and is expected to find a broad resonance in introductory courses and lectures on "2D-NMR." It is hoped that this workbook will be successful, and it is heartily recommended to all chemists as an introduction to the practical application of modern NMR spectroscopy.

Siegen, FRG, February 1989   *H. Günther*

Preface

After the first spectrometers became generally available the application of high-resolution nuclear magnetic resonance (NMR) spectroscopy to molecular structural analysis rapidly became a primary endeavor of organic chemists. The growth and development of NMR has been characterized by a series of major technological advances. In the 1950s single resonance ^1H experiments prevailed and provided basic information for a given sample on the numbers of nonequivalent nuclei, their relative shieldings, and their spin-coupled neighbors. In the early 1960s, multiple resonance techniques were introduced which permitted the extraction of more detailed information and also gave evidence of the potential of ^{13}C MR-studies. In the late 1960s, Fourier transform (FT) methods dramatically improved sensitivity, so that ^{13}C spectra could be obtained routinely. The FT technique also rendered measurements of time-dependent phenomena much more accessible. Equally important, the FT approach led to the notion of utilization of a second dimension, the potential of which was clearly recognized in the late 1970s. Through the 1980s, the implementation of these experiments has spawned new powerful methods for eliciting structural information through establishing correlations between different nuclear types and also leading to new, useful one-dimensional methods. Over this time span, advances in magnet technology provided higher and higher applied fields, increasing the sensitivity of the experiments and permitting detailed study of larger and more complex molecules. The modern high-resolution NMR spectrometer is a powerful, sophisticated system capable of providing a veritable mine of information, perhaps the most important single tool available for structural analysis.

This series of advances has perforce raised the level of sophistication required for the analysis and interpretation of the results and, while many practicing organic chemists are undoubtedly aware of instrumental capabilities, many lack direct experience with their application to real problems. Some expert guidance in the choice of specific experiments to supply the required data most reliably and, preferably, most efficiently will be welcome. Helmut Duddeck and Wolfgang Dietrich specifically address this need in the present volume. From experience gained in their own research and from organizing seminars and workshops, they have assembled 23 typical cases in this workbook to illustrate applications of modern NMR to organic structural analysis. Following clear, brief descriptions of the basic techniques, accompanied by some excellent illustrations, these cases are presented as problems to be solved by the reader. For the neophyte, strategies for approaching each case are outlined and, in the final chapter, detailed discussions of solutions for each are presented. This workbook is an excellent introduction to the practical application of modern NMR spectroscopy to structural problems and is highly recommended to all who seek guidance in the utilization of two-dimensional NMR.

London, Canada, November 1988   *J.B. Stothers*

Contents

To
Günther Snatzke
who teaches to love the architecture of three-dimensional
molecular structures.

1. Introduction

Since the early 1980s modern NMR spectroscopy – especially the two-dimensional methodology – has become an extraordinarily useful tool in the structural elucidation of unknown organic compounds. Nowadays, the latest generation spectrometers with their increasingly powerful pulse programmers, computers, and data storage devices, enable the user to perform routinely many multipulse experiments with a time expenditure no longer significantly exceeding that of most traditional techniques, as for instance, multiple selective decoupling. On the other hand, much more information can be extracted from multipulse than from conventional measurements.

Modern NMR techniques have revolutionized the structural elucidation of organic compounds and natural products. This, however, is not yet fully recognized by chemists who do not work with these methods routinely. Numerous review articles and monographs published during the last few years may give the impression that these methods are extraordinarily complicated and difficult to evaluate, thus deterring many potential users. Our experience in a number of workshops and seminars with graduate students and researchers, as well as with the routine service in our NMR laboratory, has demonstrated that in the presence of the beauty and elegance of the modern one- and two-dimensional NMR methodology, spectroscopists tend to overestimate the readiness of their "customers" to get acquainted with the underlying physical theory.

Therefore, in this book we address chemists for whom structural elucidation is an educational or occupational concern. By means of exercises taken from practice, we demonstrate that the use of spectra from multipulse NMR experiments is often straightforward and does not necessarily require insight into the underlying methodolgy and pulse sequences. For the same reason we refrain from a discussion of the physical background; the reader may find appropriate references in the bibliography. The minimal condition for successful work with this book is simply a degree of knowledge about conventional ^{1}H and ^{13}C NMR spectroscopy with which chemistry students should be familiar and that chemists can review in many textbooks or exercise collections.

Our book is fundamentally different from most other books or articles cited in the bibliography[1]. We have deliberately restricted the number of methods used to a few techniques that in the course of our daily laboratory routine, have proved executable at the spectrometer without much experimental effort and that are relatively easy to interpret. We wish to demonstrate the great potential of these few basic experiments, but without overburdening the novice with a large number of experimental variants that would be difficult to survey.

This book has been arranged so that it may serve as both a book for seminars and a self-study text for chemists who do not have access to courses. In offering a realistic picture of everyday laboratory routine, we have not attempted to plot all spectra in an optimal fashion, and, therefore, we have not tried to eliminate all artifacts. Generally, the person recording the spectra is not the same person who orders them (and often the spectroscopist does not know beforehand exactly what kind of information is to be extracted). Therefore, we want to support the reader's ability to evaluate spectra critically so that, for instance, he or she can differentiate "real" signals from artifacts. For technical reasons the spectra depicted in this book had to be reduced in size from the original plots.

Seminars on modern NMR spectroscopy have often shown that novices have a strong tendency to solve problems containing two-dimensional spectra by first and nearly exclusively evaluating the one-dimensional ^1H and ^{13}C NMR spectra and developing a structural proposal in the conventional way taught in basic courses. Later, they may try to confirm their ideas by tracing appropriate evidence from two-dimensional spectra. This approach is not essentially wrong but it is impractical and leads to a strict adherence to established structural proposals without consideration of alternatives. For instance, one often ignores the fact that a cross peak in a COSY spectrum is an unequivocal proof for the existence of a coupling and not just a probability. The observation of a signal in an NOE difference spectrum proves the spatial proximity of the respective nuclei. The novice has to learn the difference between such hard proofs and soft hints.

It is amazing to see how easy it is to establish structural fragments by simple evaluation of COSY spectra in a "jigsaw puzzle" fashion. Such an approach should always be the start of a structural elucidation. In this way, the objectivity necessary for considering all possible alternative structures is retained.

Two-dimensional spectra generally contain a wealth of information which may sometimes cause the inexperienced to become lost. The argumentation for solving a problem should therefore be structured. Preferably, one should begin with the assemblage of molecular fragments, which can later be combined into a constitution formula. Thereafter, if necessary, the stereochemistry of the compound can be investigated. In most cases this strategy leads to a quick and safe solution and an important objective of this book is to help the reader develop a feeling for this kind of approach.

However, we warn the unwary to be cautious. Two-dimensional NMR methods often give rise to artifacts, and the inexperienced tend to overinterpret such spectra. For example, the temptation to draw conclusions about the magnitude of a coupling constant from the size of a cross peak is often overwhelming. In such cases only through study, experience, and perhaps the advice of a skilled colleague can wrong conclusions be avoided.

In the choice of compounds and problems we have remained close to actual practice and offer a broad range of chemical classes representative of the chemistry for organic and natural products. The 23 exercises presented here cannot be all inclusive; because nature is unsurpassable in her variety, natural products play an important role in this book.

In Chapter 2 the experiments are discussed and explained by simple, straightforward examples. Readers without any experience in multipulse NMR spectroscopy should begin with this section.

Chapter 3 contains 23 exercises comprising signal assignments for given structures or structures known only in part, as well as for the elucidation of unknown chemical structures.

There are two levels of assistance offered by this workbook: If the reader is unable to solve the problems without assistance, there is a strategy for each exercise in its Chapter 4 section, that is, hints about how to approach the problem. The solutions themselves are described explicitly in the Chapter 5 section, and in many cases there are additional information and references. Of course, the proposed strategy is not necessarily the only possibility. With some experience the reader should be able to develop his or her own strategy independent of the descriptions in this book, which is exactly the objective we wish to achieve.

A complete signal assignment is not always necessary in order to answer the question in an exercise; occasionally, the information in the one-dimensional spectra suffices. This is intentional to show that multipulse techniques – although extremely helpful tools – are not always necessary and that even complex problems can be solved by conventional methods. We do not wish to elicit a blind and overly faithful adherence to modern NMR techniques.

Often a chemist employing modern NMR techniques faces the problem of lucidly documenting results from the spectra in a report or publication; we can offer no general rules. In Chapter 5, however, we present ways of arranging documentation in graphical and tabular form, using two particularly suitable examples (exercises 22 and 23).

In the latest NMR literature we find ab initio or a priori signal assignments, denoting spectral interpretations that are based exclusively on experimental evidence, that is, "hard" proof, and that refrain completely from the use of any empirical parameters or experience, such as chemical shifts, magnitudes of coupling constants, or substituent effects. Of course, in cases of doubt such assignments are preferable. Such a rigorous attitude, however, is coupled with a high demand for spectrometer time and familiarity with pretentious pulse programs, which not all NMR laboratories can afford and are often not required for solving a problem. Therefore, we have selected examples that allow chemists to make use of their previous experience in NMR spectroscopy.

As in our lectures and seminars it is our aim to convey something of the satisfaction that one can find in using modern NMR techniques. Fans of brainteaser problems will find a field of enjoyable activity.

---

[1] Note added in proof: Very recently, another workbook by Sanders, Constable and Hunter has appeared (cf. bibliography).

Bibliography

Reviews

Aue WP, Bartholdi E, Ernst RR (1976) Two-dimensional spectroscopy. Application to nuclear magnetic resonance. *J Chem Phys* **64**: 2229.

Bax A (1984) Two-dimensional NMR spectroscopy. *Top Carbon-13 NMR Spectrosc* **4**: 197.

Benn R, Günther H (1983) Moderne Pulsfolgen in der hochauflösenden NMR-Spektroskopie. *Angew Chem* **95**: 381; *Angew Chem Int Ed Engl* **22**: 350.

Buddrus J, Bauer H (1987) Bestimmung des Kohlenstoffgerüsts organischer Verbindungen durch Doppelquanten-kohärenz-^{13}C-NMR-Spektroskopie, die INADEQUATE-Pulsfolge. *Angew Chem* **99**: 642; *Angew Chem Int Ed Engl* **26**: 625.

Farrar TC (1987) Selective sensitivity enhancement in FT-NMR. *Anal Chem* **59**: 679 A.

Freeman R, Morris GA (1979) Two-dimensional Fourier transform in NMR. *Bull Magn Reson* **1**: 5.

Kessler H, Gehrke M, Griesinger C (1988) Zweidimensionale NMR-Spektroskopie, Grundlagen und Übersicht über die Experimente. *Angew Chem* **100**: 507; *Angew Chem Int Ed Engl* **27**: 490.

Martin GE, Zektzer AS (1988) Long-range two-dimensional heteronuclear chemical shift correlation. *Magn Reson Chem* **26**: 631.

Morris GA (1984) Pulsed methods for polarization transfer in ^{13}C NMR. *Top Carbon-13 NMR Spectrosc* **4**: 179.

Morris GA (1986) Modern NMR-techniques for structure elucidation. *Magn Reson Chem* **24**: 371.

Sadler IH (1988) The use of N.M.R. spectroscopy in the structure determination of natural products: one-dimensional methods. *Nat. Prod. Rep.* **5**: 101.

Willem R (1987) 2D NMR applied to dynamic stereochemical problems. *Progr NMR Spectrosc* **20**: 1.

Monographs

Atta-ur-Rahman (1986) Nuclear Magnetic Resonance – Basic Principles. Springer, New York.

Bax A (1982) Two-Dimensional Nuclear Magnetic Resonance in Liquids. Delft University Press, Reidel, Dordrecht.

Chandrakumar N, Subramanian S (1987) Modern Techniques in High-Resolution FT-NMR. Springer, New York.

Croasmun WR, Carlson RMK (1987) Two-Dimensional NMR Spectroscopy, Applications for Chemists and Biochemists. VCH Publishers, New York.

Derome AE (1987) Modern NMR-Techniques for Chemistry Research. Pergamon Press, Oxford.

Ernst RR, Bodenhausen G, Wokaun A (1986; 2nd ed. 1987) Principles of Nuclear Magnetic Resonance in One and Two Dimensions. Oxford University Press, Oxford.

Günther H (1983) NMR-Spektroskopie, 2nd ed. Thieme, Stuttgart, New York; (1973) NMR Spectroscopy – An Introduction. Wiley, Chichester.

Harris RK (1983) Nuclear Magnetic Resonance Spectroscopy – A Physicochemical View. Pitman, London.

Kalinowski H-O, Berger S, Braun S (1984) ^{13}C-NMR-Spektroskopie. Thieme, Stuttgart, New York; (1988) Carbon-13 NMR Spectrocopy. Wiley, Chichester.

Lambert JB, Rittner R (1987) Recent Advances in Organic NMR-Spectroscopy. Norell Press, Landisville.

Martin GE, Zektzer AS (1988) Two-Dimensional NMR-Methods for Establishing Molecular Connectivity. VCH, Weinheim.

Richards SA (1988) Laboratory Guide to Proton NMR Spectroscopy. Blackwell Scientific Publications, Oxford.

Sanders JKM, Hunter BK (1987) Modern NMR Spectroscopy, A Guide for Chemists. Oxford University Press, Oxford.

Sanders JKM, Constable EC, Hunter BK (1989) Modern NMR-Spectroscopy, A Workbook of Chemical Problems. Oxford University Press, Oxford.

Sternhell S, Field LD (1989) Analytical NMR. Wiley, Chichester.

2. Methodology

In the following sections the basic multipulse NMR techniques used in the exercises are introduced. The emphasis, however, is not on the physical description and explanation of the pulse sequences, but on the practical evaluation of the spectra and their importance in structural elucidation.

After a discussion of the advantages of high magnetic fields (Sect. 2.1) and of some one-dimensional (1D) methods useful in ^{13}C NMR spectroscopy (Sect. 2.2), NOE difference spectra are presented (Sect. 2.3). These have proved to be of extreme significance in establishing the stereochemistry of the investigated compounds.

Sections 2.4 through 2.7 deal with two-dimensional (2D) NMR spectra. There are two different kinds of 2D experiments; in the first, the so-called J-resolved (J, δ) spectra, scalar couplings (J) are displayed in the first dimension and chemical shifts (δ) in the second. The second type of experiment is with the scalar-correlated (δ, δ) spectra, in which both dimensions are associated with chemical shifts. In our NMR laboratory our routine experience (and not only ours) with correlated 2D NMR (δ, δ) spectra show them to be much broader in scope with regard to signal assignment and structural elucidation than the (J, δ) spectra. The (δ, δ) spectra provide information about the connectivity of atoms within the molecule emerging from internuclear couplings. In general, however, the magnitudes of coupling constants cannot be extracted reliably, except by using such advanced techniques as phase-sensitive COSY [1,2]. These methods, however, are not described in this book.

At the end of each section the reader can find introductory references, which, in general, are secondary literature, that is, review articles and textbooks. Our experience has shown that it is very difficult for the layman to use original publications in the correct context and to the best advantage.

All spectra (except that depicted in Fig. 2.1.1b) have been recorded using a Bruker AM-400 spectrometer operating at 400.1 MHz for ^{1}H and 100.6 MHz for ^{13}C, and equipped with a process controller, an ASPECT 3000 computer, and a CDC disk drive system (CMD, 96 MByte).

References

1. Ernst RR, Bodenhausen G, Wokaun A (1986; 2nd ed. 1987) Principles of Nuclear Magnetic Resonance in One and Two Dimensions. Oxford University Press, Oxford.
2. Kessler H, Gehrke M, Griesinger C (1988) Angew Chem **100**: 507; Angew Chem Int Ed Engl **27**: 490.

2.1 High Magnetic Fields

The development of commercially available superconducting magnets cooled by liquid helium [1], the so-called cryomagnets, has made it possible to record NMR spectra with magnetic field strengths of up to 14.1 tesla, corresponding to a proton resonance frequency of 600 MHz.

Compared to conventional electromagnets, with their maximal field strength of about 2.3 Tesla (proton resonance frequency of 100 MHz), superconducting magnets offer several advantages. First, under the influence of the higher external magnetic field, the population difference between possible spin states of NMR-active nuclei is increased, leading to a significant improvement in sensitivity. This is associated with a considerable shortening of the time required to achieve a certain signal/noise ratio. Moreover, a better resolution between the signals of nuclei with similar chemical shifts is obtained, whereas coupling constants remain unchanged since they are natural constants. For example, $\Delta\delta/J$, which is the relation between relative chemical shifts (in hertz) and the coupling constant in a two-spin system, is 3 at 80 MHz and is increased at 400 MHz by the factor $400/80 = 5$, reaching a value of 15. Thus, a strongly coupled AB spectrum at the lower field is converted to a weakly coupled AX spectrum at the higher.

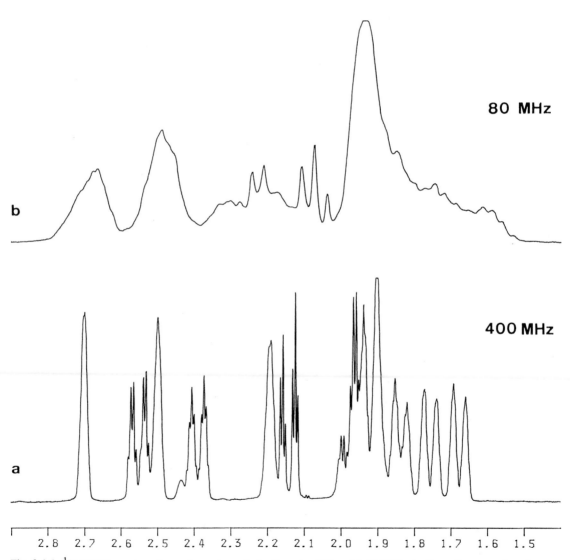

Fig. 2.1.1. ^{1}H NMR spectra of an adamantane derivative at **a** 400 and **b** 80 MHz, both on the same δ scale.

This is demonstrated impressively in Fig. 2.1.1. It is hard to believe that both ^{1}H NMR spectra belong to the same adamantane derivative; in fact, the two spectra were recorded using an identical solution. Only by comparison with the 400 MHz spectrum can it be seen that the broad peak that appears between $\delta = 2.8$ and 2.6 in Fig. 2.1.1 b does not correspond to one single proton but to an overlap of two signals that can be identified separately in Fig. 2.1.1 a, namely, that at $\delta = 2.70$ and the left part of the doublet at $\delta = 2.55$. This example demonstrates clearly that not only does a high magnetic field considerably simplify the interpretation of high-order spectra, but often it is the only way of achieving a reliable assignment of signals close to each other in the spectrum. Thus, even ^{1}H NMR spectra of such complex aliphatic molecules as steroids or triterpenoids can now be studied [2–5].

In this context, however, it should be mentioned that the predominance of dipolar relaxation processes associated with Nuclear Overhauser Effects (NOEs) may be diminished. Depending on molecular parameters, NOEs may become very small, may be suppressed, or may even become negative [6].

References

1. Günther H (1983) NMR-Spektroskopie, 2nd ed. Thieme, Stuttgart, p 264; NMR Spectroscopy – An Introduction. Wiley, Chichester, p 282.
2. Barrett MW, Farrant RD, Kirk DN, Mersh JD, Sanders JKM, Duax WL (1982) *J Chem Soc Perkin Trans* **2**: 105.
3. Schneider H-J, Buchheit U, Becker N, Schmidt G, Siehl U (1985) *J Am Chem Soc* **107**: 7027.
4. Duddeck H, Rosenbaum D, Elgamal MHA, Fayez MBE (1986) *Magn Reson Chem* **24**: 999.
5. Croasmun WR, Carlson RMK (1987) Two-Dimensional NMR Spectroscopy, Applications for Chemists and Biochemists. VCH Publishers, New York, p 387.
6. Noggle JH, Schirmer RE (1971) The Nuclear Overhauser Effect. Academic Press, New York.

2.2 One-dimensional ^{13}C NMR Spectra (DEPT)

^{13}C NMR spectra are routinely recorded under ^{1}H broadband (BB) decoupling [1]. Thus, a significant improvement of the signal/noise ratio is achieved because the signals of the insensitive ^{13}C nuclei appear as narrow singlets without any splitting due to ^{1}H,^{13}C coupling. In addition, the nuclear Overhauser effect (NOE) may enhance the signal intensities thereby as much as threefold (cf. Sect. 2.3). However, this is accompanied by a complete loss of ^{1}H,^{13}C coupling information so that, for example the number of hydrogen atoms adjacent to a carbon can no longer be determined.

In ^{1}H coupled spectra obtainable by the so-called gated decoupling technique [2, 3]; the carbon signals are split owing to the large one-bond ^{1}H,^{13}C coupling constants $^1J_{CH}$ (between 120 and 200 Hz), and doublets are observed for CH, triplets for CH_2, and quartets for CH_3 fragments, possibly over a range of several parts per million (ppm). Often these multiplets contain further fine splitting from couplings over more than one bond and may overlap severely so that an unambiguous assignment is impossible. To escape this dilemma, the so-called off-resonance spectra were invented at the beginning of routine ^{13}C NMR spectroscopy. The effect of partial ^{1}H decoupling is achieved by irradiation of a selective proton frequency near to the ^{1}H resonance range (off-resonance) [2,3]. All signal splitting due to carbon-proton couplings are reduced to such an extent that only the large one-bond couplings give rise to a relatively small amount of residual splitting, and their multiplicities indicate the number of hydrogen atoms attached to carbons.

Unfortunately, off-resonance techniques have a number of severe drawbacks. For instance, signal splittings are not always clear enough to determine multiplicities [3]. Moreover, it may be difficult to distinguish a doublet (CH) from a quartet (CH_3) signal if the signal/noise ratio is not good. The most serious disadvantage, however, becomes apparent when many ^{13}C signals exist in a narrow chemical shift range, a situation often occurring in the spectra of steroid, triterpenoids and other molecules containing many carbon atoms. In spite of the relatively small amount of residual splitting, there is still considerable signal overlap, which may easily obscure any identification of multiplets.

Modern multipulse NMR techniques offer methods that replace off-resonance experiments and are able to overcome these problems. The information – separation of ^{13}C signals according to the number of attached hydrogens – is the same; however, it does not reside in residual splittings, but in signal intensities exclusively. Peaks may be positive or negative, or they may be absent (zero intensity). This effect is obtained by the so-called *J*-modulation [4] described briefly as follows: A ^{13}C nucleus, for example, bearing only a single ^1H coupling partner displays a doublet signal; that is, according to the spin orientation of its partner, there are two spin states for this ^{13}C atom creating two magnetization vector components that differ by the value of *J* in their precession rate (Larmor frequency). After a certain delay (evolution time), without ^1H decoupling the two component vectors will be arranged so that their vector sum is either positive, negative, or zero (in the latter case the components are exactly antiparallel), leading to negative, positive, or zero-intensity signals, respectively, after a spin-echo pulse sequence [3]. It can be shown that depending on the fragment under consideration (C, CH, CH_2, or CH_3), the intensity behavior of the ^{13}C signals follows different cosine functions so that discrimination is possible [4].

Experiments based upon this principle are called *J*-modulated or *J*-coupled spin-echo measurements and are sometimes referred to under the purely descriptive acronym APT (*A*ttached *P*roton *T*est).

There is another important technique called INEPT (Insensitive Nuclei Enhanced by Polarization Transfer) [4], in which a *J*-modulation (here a sine dependence) is accompanied by a polarization transfer (PT) from the protons to coupled carbons, leading to a significant improvement in sensitivity. With this method, however, signals of quarternary carbons do not appear because the experiment is generally optimized to accomplish PT via large one-bond coupling. Nevertheless, such quaternary carbon signals can easily be detected by comparison of the INEPT spectrum with the normal ^1H broadband decoupled ^{13}C NMR spectrum.

A further improvement has been introduced by the DEPT technique (Distortionless Enhancement by Polarization Transfer) [5]. Its advantage, compared with INEPT, is a shorter pulse sequence so that during the evolution time the loss of magnetization due to transversal relaxation is less severe. Moreover, DEPT is clearly less sensitive to missettings of parameters such as pulse widths or delays (as functions of coupling constants).

The so-called spectral editing enables us to prepare DEPT spectra in such a way that only CH, CH_2, or CH_3 signals are displayed. This technique, however, requires three separate measurements. The same APT information, can also be obtained more economically by two experiments, as demonstrated in Fig. 2.2.1; this is the method of choice for all DEPT spectra in this book.

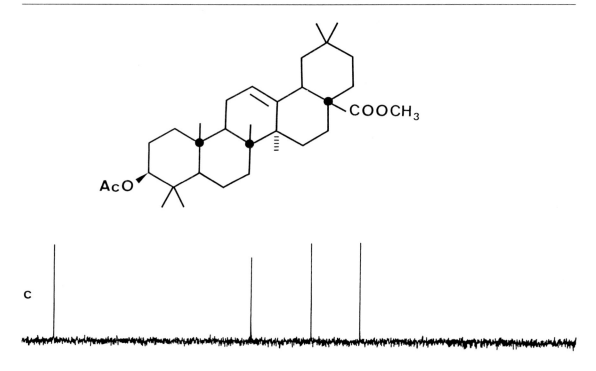

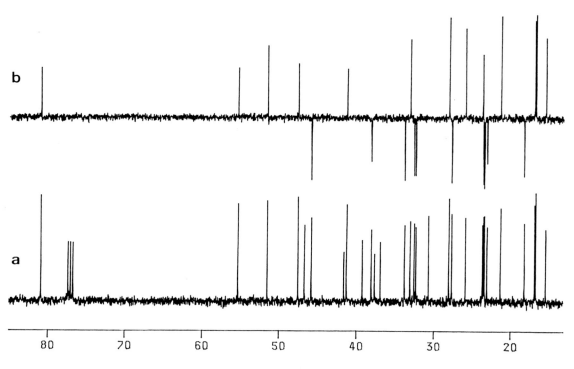

Fig. 2.2.1. ^{13}C DEPT spectra of 3-acetyloleanolic acid methyl ester, aliphatic region only: **a** broadband ^{1}H decoupled spectrum; **b** CH$_3$ and CH signals positive, CH$_2$ signals negative; **c** CH signals only.

In INEPT experiments PTs are simultaneously accomplished for all ^{1}H and ^{13}C nuclei. In general, the delays between pulses are adjusted to generate PT via one-bond ^{1}H,^{13}C couplings. An interesting variant of the INEPT pulse sequence [6] involves a "soft", that is, selective, pulse on one single proton so that ^{13}C signals appear only for those carbons that are coupled to the irradiated proton. This method is of particular interest if the delays are optimized to a long-range ^{1}H,^{13}C coupling so that quaternary carbons can be identified. This method is only feasible, however, if the signal of the irradiated proton is isolated from other signals. In Fig. 2.2.2 the application of this techniques is demonstrated using vanillin as an example.

It is apparent that, with a proton pulse on H–5 and a delay adjusted for long-range ^{1}H,^{13}C couplings of 8 Hz, only the signals of C–1 and C–3 appear with significant intensities because the respective coupling constants are the only ones meeting the 8 Hz value in a benzene ring. This example shows that it is easy to differentiate the two oxygen-bearing quaternary carbon atoms C–3 and C–4. The same information can also be obtained by two-dimensional methods (cf. Sect. 2.6), with, however, a much larger time expenditure.

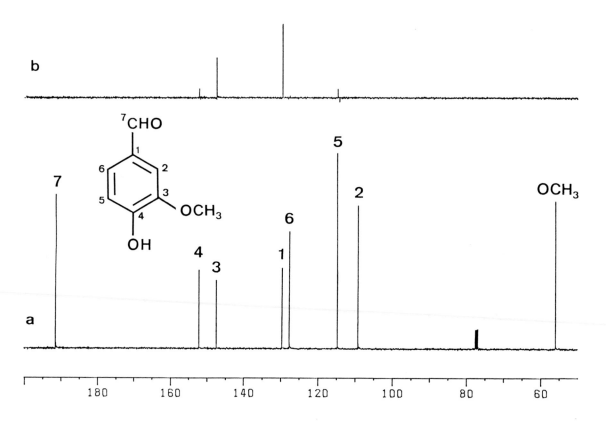

Fig. 2.2.2.a. ^{1}H broadband decoupled ^{13}C NMR spectrum of vanillin; b selective INEPT experiment with ^{1}H pulse on H–5, optimized to $^nJ_{CH} = 8$ Hz.

References

1. For modern multipulse ^{1}H broadband decoupling methods see Shaka AJ, Keeler J (1987) *Prog NMR Spectrosc* **19**: 47.
2. Kalinowski H-O, Berger S, Braun S (1984) ^{13}C-NMR-Spektroskopie. Thieme, Stuttgart, p 46.
3. Günther H (1973) NMR spectroscopy – an introduction. Wiley, Chichester, p 310, 359.
4. Benn R, Günther H (1983) *Angew Chem* **95**: 381; *Angew Chem Int Ed Engl* **22**: 350; Sanders JKM, Hunter BK (1987) Modern NMR spectroscopy – a guide for chemists. Oxford University Press, Oxford, p 69.
5. Bendall MR, Doddrell DM, Pegg DT, Hull WE (1983) High resolution multipulse NMR spectrum editing and DEPT. Bruker brochure; Derome AE (1987) Modern NMR techniques for chemistry research. Pergamon Press, Oxford, p 143.
6. Bax A (1984) *J Magn Reson* **57**: 314.

2.3 NOE Difference Spectra

The ability to measure nuclear Overhauser effects (NOEs), which enhance signal intensities, has existed for many years, and measurements have been performed using older generation continuous-wave (cw) spectrometers [1]. In the early 1960s it was shown in a double-resonance experiment that the irradiation of a proton S may lead to an up to 50% enhancement of the signal intensity of another proton I [2,3]. The most important condition for such an observation is that nucleus I be greatly relaxed by the dipolar mechanism [2,3]. It is also important that the ability of the irradiated nucleus S to influence the population difference of the transitions of nucleus I fade away with the inverse of the sixth power of the distance between both nuclei. Thus, in contrast to scalar spin-spin couplings, the appearance of NOE signal enhancements provides information about the spatial proximity of nuclei in a molecule regardless of the number of bonds between them.

Under the assumption that dipolar relaxation dominates the signal enhancement as a consequence of an NOE, it can be described by

$$\eta_{I_{max}} = \gamma_S / 2\gamma_I.$$

If both nuclei are protons, the maximal intensity gain is 50%, that is, the signal may become 1.5 times as large. If the observed nucleus is ^{13}C, the signal can be enhanced as much as threefold in an optimal case since $\gamma_{^1H} \approx 4 \cdot \gamma_{^{13}C}$. This fact has been welcomed in ^1H broadband decoupled ^{13}C NMR spectroscopy from its beginning [4] (cf. Sect. 2.2).

Measurements of NOEs using cw spectrometers have been based on intensity comparisons, and in experiments both with and without selective decoupling, the different heights of integration curves have been observed and evaluated. This method is rather limited if the NOE is small. Since the early 1980s, the so-called NOE difference technique has been used to substract free induction decays (FIDs) obtained with pulse Fourier transform (PFT) spectrometers, both with or without double resonance irradiation. These difference spectra contain signals only of such nuclei which suffer from NOE-induced intensity changes; all others are cancelled. Thus, even very small intensity differences can be reliably monitored, and there is no overlapping of uninvolved signals.

The foregoing is demonstrated in Fig. 2.3.1: The acetate of a benzodiazepinone derivative [5] has been nitrated. The question is whether the newly introduced nitro group is situated at position 7 or 8.

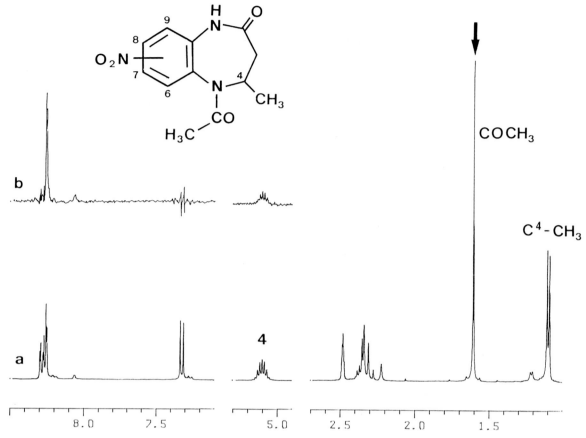

Fig. 2.3.1a and b. NOE difference experiment with a nitrated benzodiazepinone derivative, in DMSO–d_6. **a** ^{1}H NMR spectrum; **b** NOE difference spectrum with irradiation at the position of the acetoxy methyl signal (marked by the arrow).

This problem cannot be solved by establishing the H,H connectivity, since there are no detectable couplings between the aromatic and aliphatic protons.

Irradiation of the acetoxy methyl protons affords significant intensity enhancements for H–4 and for one of the aromatic protons, which apparently does not possess an ortho-positioned ^{1}H neighbor, since the signal is a narrow singlet. Owing to spatial proximity, this can only be H–6, so, it has to be concluded that the nitro group is attached to C–7. This simple experiment, requiring only a few minutes of spectrometer time, gives an answer to a question that could have been solved alternatively only by establishing C, C connectivities. This, however, would involve the use of time-consuming direct (2D INADEQUATE, Sect. 2.7) or indirect methods (COSY and COLOC, Sects. 2.4 through 2.6).

Often, in such experiments, it is possible to observe artifacts whose origin the user should know for the sake of a correct interpretation. Minor temperature or field strength deviations during the measurement can lead to residual signals with a dispersion-type appearance, even in the absence of an NOE (see, e.g., the signal at $\delta = 7.33$ in Fig. 2.3.1 b).

Occasionally intensities of partial peaks in multiplet signals are severely changed as compared with the unperturbed case; these peaks may even be negative. Such situations occur if the observed and the irradiated nuclei have a significant common coupling – for example, diastereotopic protons within a methylene group or vicinal antiperiplanar protons. This is caused by a PT between transitions with common energy levels, an effect that is successfully used in experiments involving SPT (selective population transfer) [6], INEPT, or DEPT (cf. Sect. 2.2). If the total intensity of such a multiplet, as indicated by the integration curve, is significantly different from zero, the signal can be regarded as NOE positive.

If in conformationally mobile molecules some atoms are chemically interchanging (dynamic NMR), an NOE enhancement may occur at atomic positions far from the site of the irradiated nucleus. In such cases the nucleus may have received its signal intensity enhancement in the vicinity of the irradiated nucleus and then changed its position by a fast conformational rearrangement before the original population difference in its energy levels is retained by relaxation.

It is tempting to evaluate an NOE difference experiment quantitatively in order to obtain the magnitudes of internuclear distances within a molecule, and, indeed, it is easy to extract relative intensity values (in percentages) from the computer-stored spectrum. However, the extent of a signal intensity enhancement depends on many experimental parameters, such as decoupler power, duration of decoupler irradiation, presence of relaxation mechanisms other than dipolar, and correlation times of the molecule. Therefore, a quantitative evaluation should be restricted, if made at all, to molecules very similar in structure, to spectra obtained under identical external conditions, and to experiments for which the signal enhancements obtained can be calibrated using known interatomic distances. A semiquantitative interpretation, however (signals indicated as strong, medium, weak, or absent), is significant, often very useful, and in most cases sufficient.

Rarely found in the literature are heteronuclear variants of NOE difference experiments in which protons are irradiated selectively and signal enhancements for ^{13}C nuclei are observed. The main problem is that carbon nuclei are very efficiently relaxed by their own directly attached protons so that NOEs from other protons farther away cannot produce additional significant signal enhancements. Thus, heteronuclear NOE experiments are largely restricted to the observation of signals belonging to quarternary carbons.

The preceding is demonstrated in Fig. 2.3.2, showing the differentiation between the aliphatic quarternary C–1 and C–3 in fenchone. If H–4 is irradiated, a significant NOE is observed for C–3 but not for C–1. Among the hydrogen-bearing carbons only C–4 directly attached to H–4[1]and, to a smaller extent, C–5 are affected.

Spatial proximities can also be derived from 2D (the so-called NOESY) experiments. These spectra are very similar to H,H COSY spectra (Sect. 2.4); the cross peaks, however, do not indicate scalar (through bond) but rather, dipolar (through space) couplings. It has been shown [7] that the NOESY

---

[1]The fact that the C–4 signal appears at all is surprising, since only the H–4 protons that are attached to ^{12}C–4 nuclei have been irradiated. Those at ^{13}C–4 atoms are represented by the ^{13}C satellites, which are approximately 60 to 70 Hz away from each side of the main signal. Probably, the decoupler power was strong enough to affect not only the main signal but also these satellites.

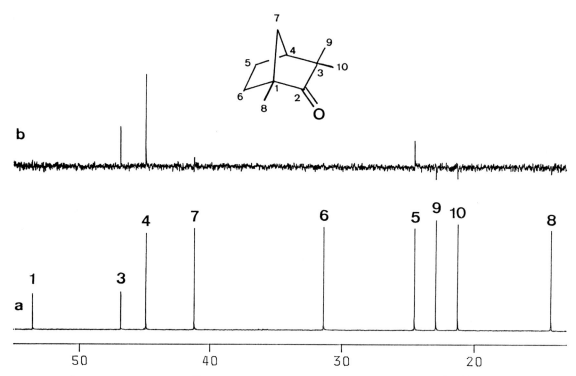

Fig. 2.3.2a. [1]H broadband decoupled ^{13}C NMR spectrum of fenchone; **b** heteronuclear (^{1}H,[13]C) NOE difference spectrum with H–4 irradiated.

technique is preferably applied to molecules with high molecular weights, for example, biopolymers. For smaller compounds like those considered in this book, the NOE difference spectroscopy is more suitable [7]. In addition, NOE difference is often executed with shorter spectrometer time.

References

1. Von Philipsborn W (1971) *Angew Chem* **83**: 470; *Angew Chem Int Ed Engl* **10**: 472.
2. Noggle JH, Schirmer RE (1971) The Nuclear Overhauser Effect. Academic Press, New York.
3. Günther H (1973) NMR spectroscopy – an introduction. Wiley, Chichester, p 299.
4. Kalinowski H-O, Berger S, Braun S (1984) ^{13}C-NMR-Spektroskopie. Thieme, Stuttgart, pp 44, 566.
5. Ried W, Urlass G (1953) *Chem Ber* **86**: 1101; Snatzke G, Malik F, Duddeck H (1987)(to be published).
6. Martin ML, Martin GJ, Delpuech J-J (1980) Practical NMR spectroscopy. Heyden, London, p 222; Derome AE (1987) Modern NMR techniques for chemistry research. Pergamon Press, Oxford, p 130.
7. Ernst RR, Bodenhausen G, Wokaun A (1986; 2nd ed. 1987) Principles of nuclear magnetic resonance in one and two dimensions. Oxford University Press, p 516.

2.4 ^{1}H,^{1}H Correlated (H,H COSY) 2D NMR Spectra

One of the most important 2D techniques is H,H COSY, the spectra of which display ^{1}H chemical shifts in both dimensions. H,H COSY spectra are obtained by a series of individual measurements that differ from each other by an incrementally changed delay (t_1) between two 90° pulses [1,2]. Thus interferograms are obtained in the time domain t_2 (free induction decays or FIDs) and are differently modulated because of the variable t_1 time. It is important to note that by this procedure ^{1}H chemical shift information is present not only in the FIDs themselves, but also in their modulation. In a first step the FIDs are Fourier transformed (as is usual in 1D NMR spectroscopy) to create spectra in the frequency domain F_2 [1–3]. A second Fourier transformation in the t_1 direction provides the second frequency dimension (F_1) of the 2D NMR spectra [1,2].

One-dimensional NMR spectra are, of course, two-dimensional, the second dimension being the signal intensity. Correspondingly, 2D NMR spectra are three-dimensional. Therefore, reproducing such spectra on paper is a problem because the spectra have to be reduced by one dimension. There are two principal ways of achieving this: Either the spectrum is depicted in a perspective view, or the intensity dimension is eliminated and the lost information restored, at least in part, by the introduction of contour lines like in a topological map.

In the first case one obtains the so-called stacked plot (Fig. 2.4.1) which contains the complete intensity information and catches one's eye because of its appearance. Unfortunately, stacked plots suffer from several drawbacks. First, an interpretation is hampered by the perspective distortion. Second, it cannot be determined whether small signals are hidden behind larger ones owing to the "whitewashing" of peaks. In case of doubt a second plot is necessary from a different angle of perspective. Third, the plotting of such a spectrum is rather time consuming and may take one hour or longer.

The second alternative is the so-called contour plot. As already mentioned, intensity information is partly lost; in cases of doubt, however, it can be regained by plotting traces in any desired direction. The contour lines are obtained by intersecting the 3D spectrum with planes parallel to the F_1, F_2 plane at consecutive heights. The lowest level of the planes and their number determine how much intensity information is restored. If the lowest level is too low, many noise peaks will appear, obscuring the real signals. If it is too high, there is the risk that small but real peaks will be ignored. The main advantages of contour plots are that they are very easy to survey and signal hiding, as in stacked plots, is impossible. Furthermore, there is no perspective distortion, and the actual plotting takes only a few minutes.

In theory, H,H COSY spectra are symmetrical with respect to the diagonal, since both frequency domains contain the same ^{1}H chemical shift information. In practice, however, such symmetry is seldom observed because the digital resolution is quite different in both dimensions (cf. the two projections in Fig. 2.4.2). Moreover, artifacts without any symmetrical counterpart frequently exist (cf. Figs. 2.4.1a and 2.4.2). Such artifacts originate in incorrect pulse widths, too short relaxation delays, longitudinal relaxation during the evolution time t_1, and other experimental imperfections. In order to eliminate these imperfections, a mathematical algorithm, the so-called symmetrization algorithm, can be applied. This procedure compares the memories of data points that are symmetrical pairwise and uses the lower one for both, thereby eliminating all signals that do not possess a symmetrical counterpart (cf. Figs. 2.4.2 and 2.4.3).

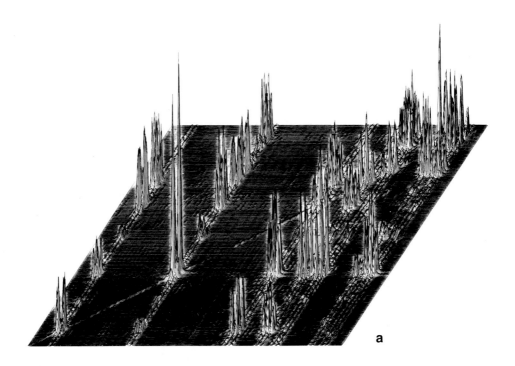

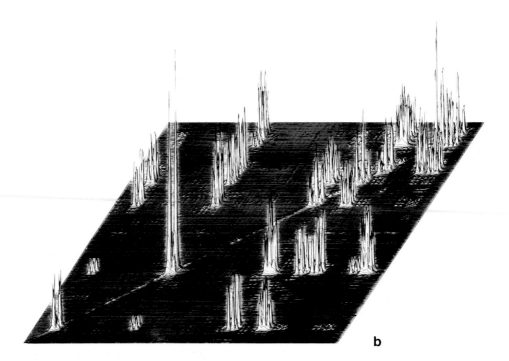

Fig. 2.4.1. Stacked plot of an H,H COSY spectrum of N-methyl-benzoisocarbostyril [4], aromatic region only; (a) not symmetrized; (b) symmetrized.

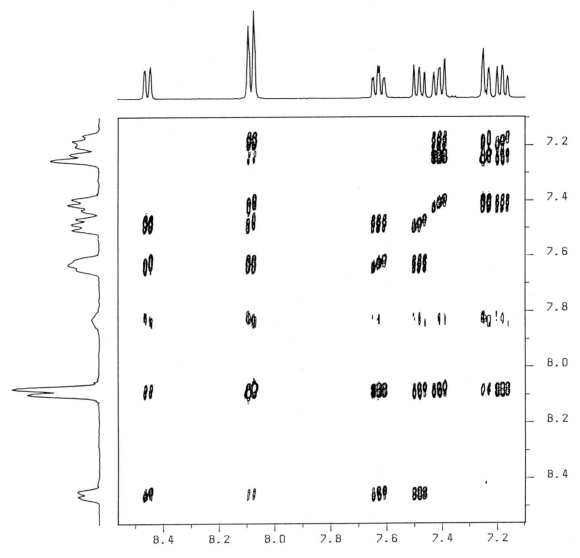

Fig. 2.4.2. Contour plot of an H,H COSY spectrum of N-methyl-benzoisocarbostyril [4], aromatic region only, not symmetrized.

Symmetrization of the 2D data matrix facilitates interpretation of the spectra and, in addition, leads to an improvement in the signal/noise ratio by a factor of $\sqrt{2}$, a welcome extra result especially if only a small quantity of substance is available. It should not be ignored, however, that symmetrization may also have disadvantages. Artifacts that by chance have a symmetrical counterpart will not be removed and will give the impression that they are real. In practice, however, we have found that the advantages predominate, and, therefore all H,H COSY spectra in this book are symmetrized.

Two basically different types of signals appear in H,H COSY spectra. Those at the diagonal (diagonal peaks) represent the original spectrum, as obtained in a 1D experiment. The off-diagonal signals are the so-called cross peaks, which prove the existence of scalar (through-bond) couplings

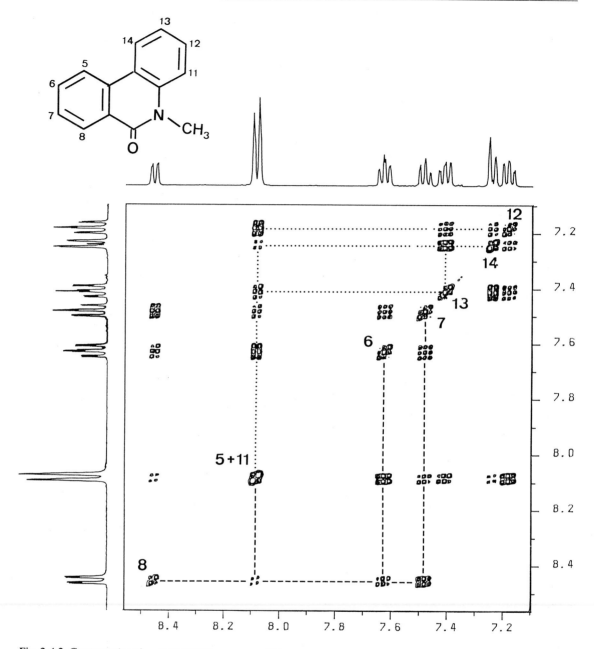

Fig. 2.4.3. Contour plot of an H,H COSY spectrum of N-methyl-benzoisocarbostyril, aromatic region only, symmetrized; ^{1}H signal assignment taken from [4].

between nuclei. The corresponding coupling partners can be found by drawing horizontal and vertical lines starting at the cross peak until the diagonal is intersected, and these positions are the signals of the coupling partners. Owing to the symmetry of the spectrum, this procedure can be performed in either the upper left or the lower right triangle.

An H,H COSY measurement can be regarded as equivalent to a series of selective decouplings during which all chemically different protons are decoupled consecutively. Such experiments, however, are laborious and, because of signal overlap, often inconclusive, so the two-dimensional technique is clearly superior. It should be noted, however, that by no means can simple H,H COSY spectra replace decoupling measurements in which ^{1}H multiplets are to be simplified for identifying coupling constants.

The evaluation of an H,H COSY spectrum is explained in the following using N-methylbenzoisocarbostyril (Fig. 2.4.3); the ^{1}H signal assignment is taken from [4]. The proton absorbing at the highest frequency ($\delta = 8.44$) is H–8 because it is in the peri-position with respect to the carbonyl group. This signal has three cross peaks, which are identified by the horizontal dashed line in the lower triangle of the diagram. Each is the starting point of vertical dashed lines meeting the diagonal at the signals of the protons coupled to H–8 (H–5, H–6, and H–7). The signal at $\delta = 8.08$ corresponds to two accidentally isochronous protons, and the second signal must be a doublet, that is, it has only one ortho neighbor. Indeed, it is H–11 [4]. In the upper triangle of the diagram, dotted lines can be seen indicating how the coupling partners of H–11, namely, H–12, H–13, and H–14, are found.

It is interesting to see that the signal multiplicities of coupling nuclei are reflected in the cross peaks. For instance, the cross peak connecting H–5 and H–8 displays $2 \times 2 = 4$ partial peaks, and that connecting H–6 and H–7 as many as $3 \times 3 = 9$.

If the signals of coupling partners are close to each other, that is, if they have very similar chemical shifts, the corresponding cross peak is located very near to the diagonal and may be obscured by overlap of the diagonal peaks. In such case there are variants and improvements of the COSY pulse sequence to alleviate the situation. In the so-called COSY45 variant the second pulse is not a 90° but a 45° pulse, decreasing the extension of the diagonal peaks [2]. Therefore, all H,H COSY spectra in this book have been recorded using the COSY45 pulse sequence. This technique offers another advantage. If the digital resolution is good enough, it may be possible to extract the sign of the coupling constant from the unsymmetrical form of the cross peak, as can be observed in Fig. 2.4.4 for the cross peak connecting H–3 and each of the H–10 (or H–7 and each of the H–10). The cross peaks are not symmetrical, and the dashed line indicates the "longest diagonal" within the peak. The direction of this dashed line – here a negative slope – allows us to conclude that the coupling constant is positive; indeed, it is a vicinal $^3J_{HH}$. Correspondingly, positive slopes are sometimes observed when the coupling constants are negative, for instance, $^2J_{HH}$ values.

Further improvements can be achieved in digital resolution, and even coupling constant values may be extracted from the spectra of the so-called double-quantum filtered, phase-sensitive COSY spectra [5].

H,H COSY spectra from samples with a substrate concentration of about 0.3 to 0.5 mM or more can be obtained in relatively short spectrometer time. If one is mainly interested in H,H connectivity rather than high resolution, a total recording time of about one hour should be sufficient. Of course, the minimal substance requirement also depends on a number of additional factors such as ^{1}H relaxation times and the sensitivity of the spectrometer: as a general rule, the higher the external field, the better the sensitivity.

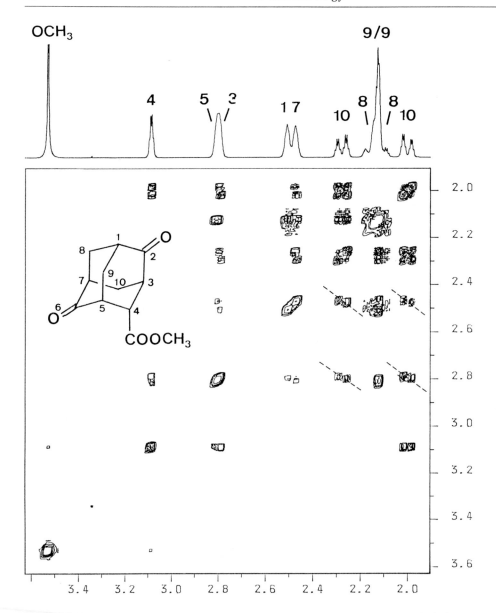

Fig. 2.4.4. H,H COSY spectrum (COSY 45) of 4-methoxycarbonyl-adamantane-2,6-dione.

Recently, the potential of COSY spectroscopy has been expanded by the introduction of the so-called RELAY technique [1,5]. To understand its principle, let us imagine a three-spin proton system (A · · · B · · · C) in which A and B, as well as B and C, are coupled pairwise, but A and C are not. The homonuclear RELAY experiment creates a PT from proton A to proton B, which is the relay, and passes it on to C. Thus, in an H,H,H RELAY spectrum a cross peak connecting A and C is observed, although these nuclei do not have a common coupling. A comparison with the respective H,H COSY spectrum that does not display such a peak can provide further H,H connectivity information.

References

1. Benn R, Günther H (1983) *Angew Chem* **95**: 381; *Angew Chem Int Ed Engl* **22**: 350.
2. Derome AE (1987) Modern NMR techniques for chemistry research. Pergamon Press, Oxford.
3. Sanders JKM, Hunter BK (1987) Modern NMR spectroscopy − a guide for chemists. Oxford University Press, Oxford.
4. Duddeck H, Kaiser M (1985) *Spectrochim Acta* **41A**: 913.
5. Ernst RR, Bodenhausen G, Wokaun A (1986; 2nd ed. 1987) Principles of nuclear magnetic resonance in one and two dimensions. Oxford University Press, Oxford, p 434.

2.5 ^{1}H,^{13}C Correlated (H,C COSY) 2D NMR Spectra

The H,C COSY measurement is extremely important, since it connects ^{1}H signals in the F_1 dimension with ^{13}C signals in F_2 [1−3]. Similarly, as in the homonuclear case (Sect. 2.4), the 1D equivalent of H,C COSY is a series of decoupling techniques in which each proton is irradiated selectively [4]. The problem, however, in such 1D experiments is that in a ^{1}H NMR spectrum it is necessary to irradiate the ^{13}C satellites rather than the main proton signals. These satellites are doublets and are many hertz apart from each other because of the large one-bond ^{1}H,^{13}C couplings ($^1J_{CH}$). Therefore, in contrast to homonuclear decoupling experiments, considerably larger decoupling power has to be applied, leading to unwanted off-resonance effects for other signals (cf. Sect. 2.2). Such effects are particularly severe when ^{1}H signals are so close that their ^{13}C satellite doublets partially overlap.

As for H,H COSY, the graphic representation of H,C COSY spectra can be either a stacked or a contour plot (Sect. 2.4), and again the contour plot is preferred for the same reasons. Since in this case the chemical shift information is different for the two dimensions, there is, of course, no symmetry. Consistently throughout this book, the ^{1}H dimension (F_1) is plotted vertically and the ^{13}C dimension (F_2) horizontally in all heterocorrelated 2D NMR spectra (see also COLOC, Sect. 2.6).

The cross peaks in Fig. 2.5.1 prove which hydrogen atoms are directy attached to which carbons. Note the signals of the methylene groups C−8, C−9, and C−10. For C−10 there are two distinct signals, corresponding to the two diastereotopic and anisochronous protons; the chemical shift difference is nearly 0.3 ppm. The δ values of the two H−8 are still discernible whereas for the two H−9 a chemical shift difference can no longer be detected. This shows that by no means can an H,C COSY replace a DEPT spectrum as an APT experiment (cf. Sect. 2.2).

In H,C COSY spectra only signals for CH$_n$ fragments with $n \geq 1$ are visible, that is, there is no information about quaternary carbons. The reason is a delay time in the pulse sequence that is proportional to the inverse of ^{1}H,^{13}C coupling constants and that is calibrated for the large one-bond coupling constants ($^1J_{CH}$ = 120−200 Hz). This delay time can be optimized for smaller couplings. It then, however, has to be increased to such an extent (several hundred milliseconds) that measurements are only feasible if the transversal relaxation time of the protons (T_2^*) is relatively large. Otherwise, the magnetization will have decayed more or less at the end of the pulse sequence (i.e., when the FID is to be sampled), and the experiment will be very insensitive.

All H,C COSY spectra in this book have been recorded using a technique affording a quasi-^{1}H decoupled spectrum in the ^{1}H dimension. This is advantageous because the ^{1}H signals have a higher

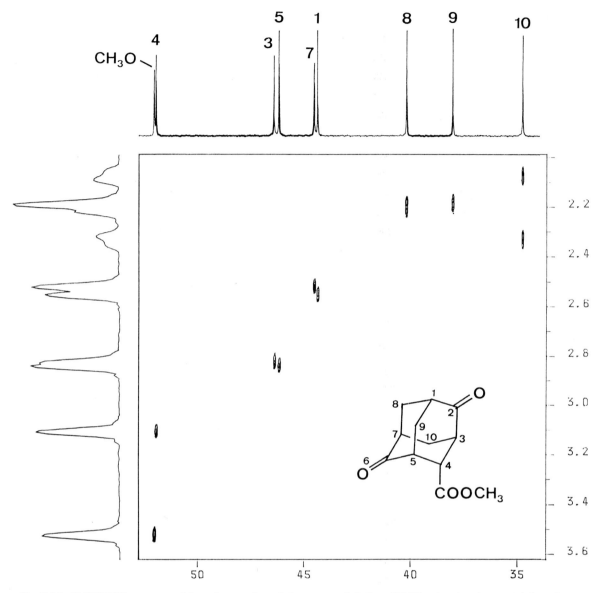

Fig. 2.5.1. H,C COSY spectrum of 4-methoxycarbonyladamantane-2,6-dione [5]. The signal assignment is based on an H,H COSY experiment (Sect. 2.4).

F_1 dispersion that improves the signal/noise ratio. The signals of methylene groups with two diastereotopic, and hence anisochronous, protons (e.g., C–10 in Fig. 2.5.1), however, tend to give small signals because the total intensity is distributed in two parts. Moreover, these signals often display a splitting, or at least a broadening, since the relatively large geminal coupling between the diastereotopic protons remains visible. In addition, an artificial signal may appear exactly midway between the two partial signals.

For all H,C COSY spectra in this book the ^{13}C spectra at the top are not projections of the actual 2D spectra. They are instead the original 1D ^{13}C NMR spectra in order that the signals of quaternary carbons, which would otherwise be absent from the projection, may also be displayed.

Provided that sufficient material is available, that is, that the concentration in the sample is 0.5 M or more, the time demand for a heteronuclear H,C COSY experiment is similar to that for a homonuclear H,H COSY spectrum. If it is possible to obtain a 1D broadband decoupled ^{13}C NMR spectrum of a given sample within a few minutes with a reasonable signal/noise ratio, an H,C COSY spectrum can be obtained in less than one hour.

The RELAY technique already mentioned in Sect. 2.4 can be applied in a heteronuclear experiment as well [6]. A proton H^A can transfer polarization via a coupled proton H^B (relay nucleus) to a carbon C^B directly attached to H^B. Thus, it is possible to monitor H,C connectivities in a molecule, which otherwise could only be detected – if at all – by an H,C COSY experiment with delay times optimized to a long-range ^{1}H,^{13}C coupling constant or by a COLOC experiment (Sect. 2.6). Moreover, this experiment is successful even if there is no significant coupling between H^A and C^B.

References

1. Benn R, Günther H (1983) *Angew Chem* **95**:381; *Angew Chem Int Ed Engl* **22**:350.
2. Derome AE (1987) Modern NMR techniques for chemistry research. Pergamon Press, Oxford.
3. Sanders JKM, Hunter BK (1987) Modern NMR spectroscopy – a guide for chemists. Oxford University Press, Oxford.
4. Günther H (1973) NMR spectroscopy – an introduction. Pergamon Press.
5. Duddeck H (1983) *Tetrahedron* **39**:1365.
6. Ernst RR, Bodenhausen G, Wokaun A (1986; 2nd ed 1987) Principles of nuclear magnetic resonance in one and two dimensions. Oxford University Press, Oxford, p 479.

2.6 COLOC Spectra

H,C COSY is largely limited to one-bond couplings and becomes insensitive if it is optimized to a small long-range ^{1}H,^{13}C coupling constant. In such case the corresponding delay times have to be quite long, so, owing to transversal relaxation, the magnetization is more or less decayed at the end of the pulse sequence before it can be monitored. A pulse sequence offering a way out of this dilemma is COLOC (Correlation Spectroscopy via Long-Range Couplings) [1]. This technique is particularly suitable if the molecule contains quarternary carbons. Since ^{1}H,^{1}H couplings influence signal intensities of COLOC peaks in a way not easy to predict, COLOC should preferably be applied to molecules or molecular fragments bearing only few hydrogen atoms.

In Fig. 2.6.1 the COLOC spectrum of vanillin is shown; the signal assignment is based on ^{13}C NMR data taken from the literature [2], from a selected INEPT experiment (Fig. 2.2.2), and from an H,C COSY spectrum. In general, peaks originating from one-bond ^{1}H,^{13}C couplings can be found in COLOC spectra as well; in Fig. 2.6.1 these have been noted by circles. In order to separate these peaks accurately from those representing long-range couplings, it is always advisable to compare the COLOC with the H,C COSY spectrum of the same compound. In the COLOC spectrum of vanillin the

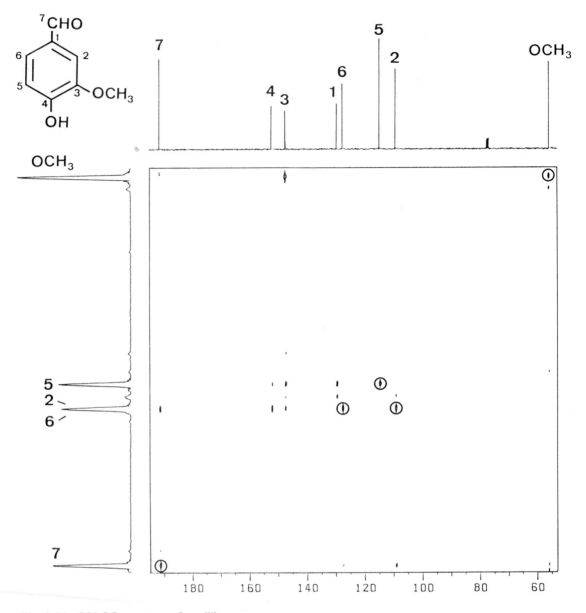

Fig. 2.6.1. COLOC spectrum of vanillin.

pulse sequence parameters have been adjusted so that ^{1}H,^{13}C coupling constants in the range of 4 to 8 Hz will give rise to significant signals. This is the typical range for three-bond couplings in coplanar atomic arrangements, and, indeed, there are a number of corresponding peaks, for instance, C–1/H–5, C–2/H–7, C–3/H–5, C–4/H–2, C–4/H–6, C–7/H–2, and C–7/H–6. The signal connecting C–3 with the methoxy protons is of special interest since it proves that the methoxy group is attached to C–3 and the hydroxy group to C–4, and not vice versa.

References

1. Kessler H, Griesinger C, Zarbock J, Looslie HR (1984) *J Magn Reson* **57**: 331.
2. Breitmaier E, Voelter W (3rd ed 1987) ^{13}C NMR spectroscopy. Verlag Chemie, Weinheim, New York.

2.7 2D ^{13}C,^{13}C (C,C COSY) INADEQUATE Spectra

Structural elucidation of an unknown organic compound or a natural product implies establishing the connectivity of the atoms in the carbon skeleton. The methods described in the previous sections only achieve this goal indirectly. For example, first by evaluation of the H,H COSY spectrum the connectivity of the protons is established, then, in a second step, H,C COSY and COLOC experiments show to which carbons these protons are bonded so that the C,C connectivity is finally obtained.

Of course, it would be much more elegant to arrive at the C,C connectivity directly from ^{13}C,^{13}C couplings. Since, however, only every ninetieth carbon is a ^{13}C isotope, only one in about 8000 molecules contains two ^{13}C nuclei in two ascertained positions. Thus, the sensitivity of such a measurement is extremely low, even at high concentrations. Moreover, the signals from ^{13}C satellites in the ^{13}C NMR spectrum can easily be overlapped by the main signal arising from molecules containing only a single ^{13}C nucleus. In addition, rotation sidebands and peaks from traces of impurities may obscure the identification of the ^{13}C satellites.

These problems can be overcome by the INADEQUATE technique (Incredible Natural Abundance DoublE QUAntum Transfer Experiment) [1-3], which suppresses the main signals so that only ^{13}C satellites appear in the spectrum; this technique also removes rotation sidebands and signals from impurities. Thus, in a 1D INADEQUATE spectrum there are one or more doublets for each carbon, according to its topological position, from which the ^{13}C,^{13}C coupling constant(s) can be taken.

Unfortunately, however, one-bond ^{13}C,^{13}C coupling constants are very uniform in CH$_x$ fragments without further electronegative substituents ($^1J_{CC}$ = 30 to 40 Hz), so it is very often difficult or even impossible to establish a positive C,C connectivity from these data alone. In cyclooctanol (Fig. 2.7.1) all $^1J_{CC}$ values are between 34.2 and 34.5 Hz; the only exception is the coupling between C-1 and C-2, which is larger (37.5 Hz) because C-1 is substituted by the hydroxy group. It can be seen, even in such a simple example, that the determination of C,C connectivities is not possible from a 1D experiment alone.

Here the transition from 1D to 2D spectroscopy is very helpful. It is possible to obtain 2D INADEQUATE (C,C COSY) spectra (see Fig. 2.7.2) that resemble the H,H COSY spectra introduced in Sect. 2.4. The only difference is that the diagonal peaks appearing in the H,H COSY spectra are absent from those of C,C COSY because diagonal peaks represent peaks from ^{13}C atoms having ^{12}C neighbors, and these have been filtered out by the INADEQUATE technique. In an evaluation similar to that of the H,H COSY spectra, the connectivity of all carbons in cyclooctanol can be obtained starting with the obvious signal to assign, namely, C-1. Below the corresponding doublet in the trace above the 2D plot, there is a cross peak leading to C-2 if we follow the horizontal dashed line until it intersects the diagonal (dotted line). From here a vertical dashed line identifies the C-2 signal. For C-2 there is

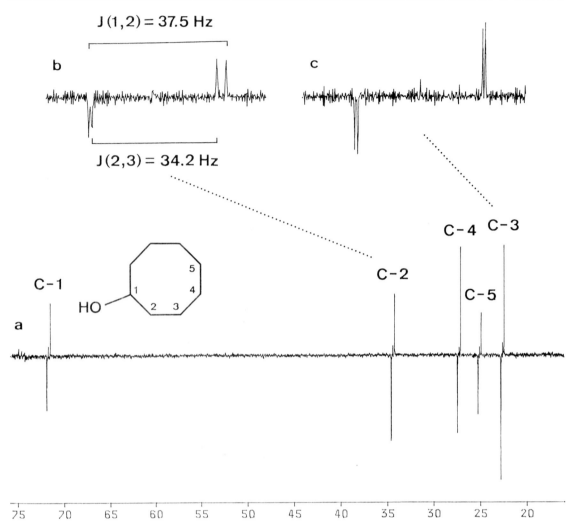

Fig. 2.7.1a. One–dimensional INADEQUATE spectrum of cyclooctanol; **b** signal of C–2 expanded; **c** signal of C–3 expanded. In these expansions the digital resolution was increased by a mathematical treatment (resolution enhancement) of the time-dependent primary spectrum (FID). With the INADEQUATE pulse sequence, the partial signals of the doublets have different signs (no refocusing). In **c** the appearance of two doublets does not indicate that the two couplings (C–3/C–2 and C–3/C–4) are of different magnitudes. Rather, the C–3 signals in the two isotopomers ($C_8H_{16}O-2,3^{13}C_2$ and $C_8H_{16}O-3,4^{13}C_2$) do not coincide, that is, here a ^{13}C neighbor isotope effect is observed.

another cross peak from which a new horizontal dashed line is drawn marking another intersection with the diagonal. From there we end up at C–3. The signals of C–4 and C–5 can be identified analogously.

Like any other 2D method, this technique requires a series of individual FID measurements. Thus, even with a very large amount of material available for the sample solution, a 2D INADEQUATE experiment is very time-consuming; often it requires one or two days of spectrometer time. Conse-

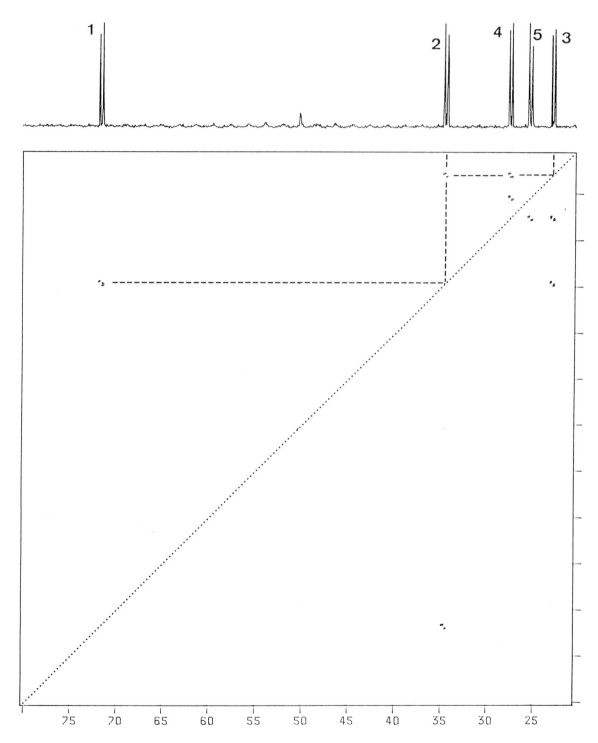

Fig. 2.7.2. 2D INADEQUATE spectrum of cyclooctanol; the 1D spectrum on the top of the plot is the projection, not the normal ^1H broad-band decoupled ^{13}C NMR spectrum.

quently, such measurements are performed only when indirect methods for establishing C,C connectivities fail.

The sensitivity of the experiment can be enhanced [3] by a combination of PT (for instance INEPT or DEPT) and INADEQUATE pulse sequences or by mathematical treatment. With further improvements, the INADEQUATE technique will probably become one of the most valuable methods in the arsenal of NMR spectroscopists.

References

1. Benn R, Günther H (1983) *Angew Chem* **95**: 381; *Angew Chem Int Ed Engl* **22**: 350.
2. Derome A E (1987) Modern NMR Techniques for Chemistry Research. Pergamon Press, Oxford, p 234.
3. Buddrus J, Bauer H (1987) *Angew Chem* **99**: 642; *Angew Chem Int Ed Engl* **26**: 625.

3. Exercises

Many of the exercises in this book contain additional information about aggregate states, characteristic IR bands, molecular formulas obtainable from high-resolution mass spectra, and so forth. Often the constitution formula is given, and sometimes statements about the configuration are offered. This closely follows actual practice, since prior to taking an NMR measurement it is normal laboratory procedure to have on hand the chemical history of the compound and information derived from other spectroscopic methods.

One of the main sources of information in NMR spectroscopy is still the 1D ^1H spectrum, whose significance is further increased by the availability of high magnetic fields (^1H resonance frequencies up to 600 MHz). Spectra recorded at lower fields (resonance frequencies of 60 to 100 MHz) are often of high order and not accessible to direct evaluation. Many of these can be converted to first-order spectra if higher fields ($\geq$ 200 MHz) are employed so that coupling constants can be read immediately from signal splittings. Even when the whole spectrum or parts of it are of high order, it is often possible to find evidence for the existence of large and/or small couplings merely from a signal's splitting pattern, even though magnitudes of the coupling constants cannot be determined directly (cf. Sect. 2.1). For instance, the form of aromatic proton signals can provide information as to whether there are ortho-positioned protons and, if so, how many, since the corresponding three-bond $^1H,^1H$ coupling constants are relatively large (6 to 9 Hz).

For almost every exercise we provide a basic set of spectra comprising the 1D ^1H and ^{13}C NMR spectra, and including DEPT, as well as the 2D H,H COSY (homonuclear) and H,C COSY (heteronuclear) spectra. Moreover, a certain standard of representation is used. The H,H COSY spectrum is always positioned on a left-hand page and the H,C COSY spectrum on a right-hand page. The proton dimension of the H,C COSY spectrum (vertical) corresponds in its chemical shift (δ) range exactly to the H,H COSY spectrum so that both plots can be easily compared. A proton signal identified in one of the two spectra can be traced in the other by simply drawing a horizontal line from its peak to the other spectrum. For the sake of better comparison, the trace at the top of the H,H COSY plot is not the projection from the experiment, but the original 1D ^1H NMR spectrum. Similarly, the horizontal trace at the top of the H,C COSY plot is the original ^1H broadband decoupled ^{13}C NMR spectrum; the vertical trace (^1H), however, is the projection of the 2D spectrum. In addition, the two ^{13}C DEPT spectra are depicted below the H,C COSY plot.

The ^{13}C chemical shifts cannot be determined exactly (i.e., with a precision better than $\pm$ 1 ppm) from the ^{13}C NMR spectra. Generally, such precision is not necessary for solving the problem; the exact values can be found in the sections entitled "Chapter 5" (solutions) at the end of each exercise.

Signals of quarternary carbons appear in the ^1H broadband decoupled ^{13}C NMR spectra, but not in those of DEPT and H,C COSY. If such signals are out of range of the H,C COSY spectrum, they, of course, do not appear in the ^{13}C NMR spectrum at the top of H,C COSY either. In such case their chemical shifts are noted in the captions with an additional note "C".

In the figures a ^{1}H NMR spectrum may be divided into several fractions for the sake of better recognition. The fractions always have the same chemical shift (in hertz/centimeters) and intensity scale. Thus, coupling constants can easily be determined (1 ppm corresponds to 400 Hz), and their splittings and intensities compared within different fractions.

For the sake of greater clarity, we have omitted integration steps in the ^{1}H NMR spectra. In most cases the number of hydrogens corresponding to a given signal is obvious; if there is any doubt, notes are provided, (e.g., "2H" for two protons).

When not otherwise indicated, deuterated chloroform was used as the solvent and the concentrations were, in general, between 0.1 and 0.5 M. The chemical shifts refer to the δ scale. Reference signals were CHCl$_3$ ($\delta = 7.24$) for ^{1}H and the central peak of CDCl$_3$ ($\delta = 77.0$) for ^{13}C. Samples with other solvents were referenced analogously.

Exercise 1

The NMR spectra depicted in Figs. 3.1.1 through Fig. 3.1.3 belong to a pleasant smelling liquid. What is the structure of this compound (1), which is a natural product with the molecular formula $C_{13}H_{20}O$?

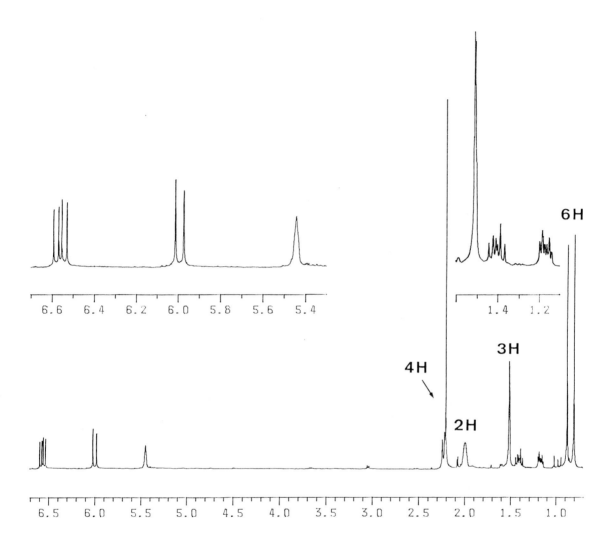

Fig. 3.1.1. ^{1}H NMR spectrum of 1.

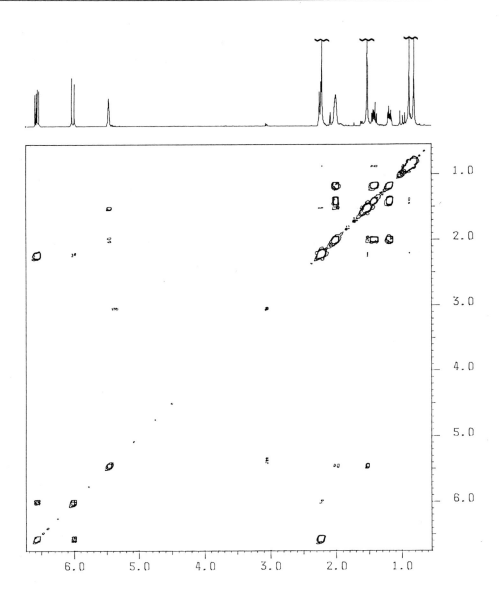

Fig. 3.1.2. H,H COSY spectrum of **1**.

Fig. 3.1.3. H,C COSY and DEPT spectra of **1**; the ^{13}C NMR spectrum contains an additional signal at $\delta = 198.4$ (C).

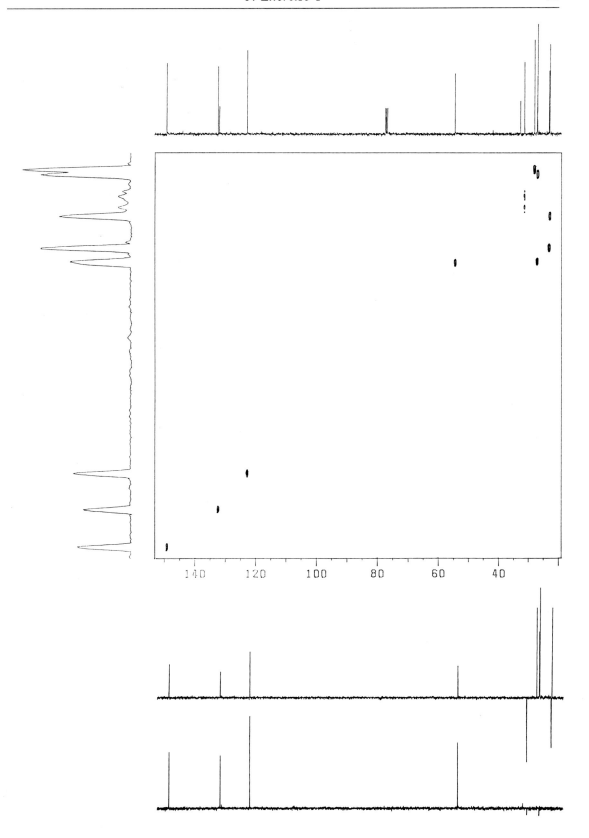

Exercise 2

Compound **2** has been isolated from the plant *Aristolochia argentina*, and its molecular weight is $C_{12}H_{18}O_2$. What is the structure of **2**?

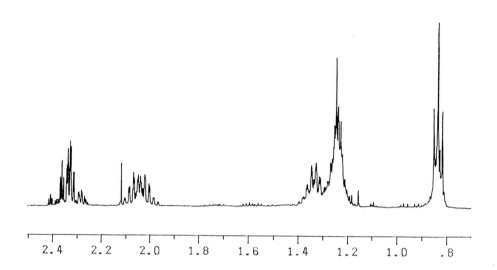

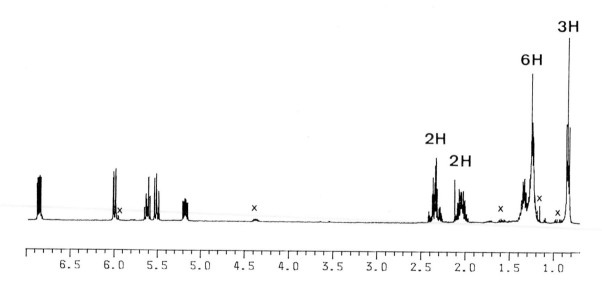

Fig. 3.2.1. ^1H NMR spectrum of **2**; the little peaks marked by "x" belong to an impurity.

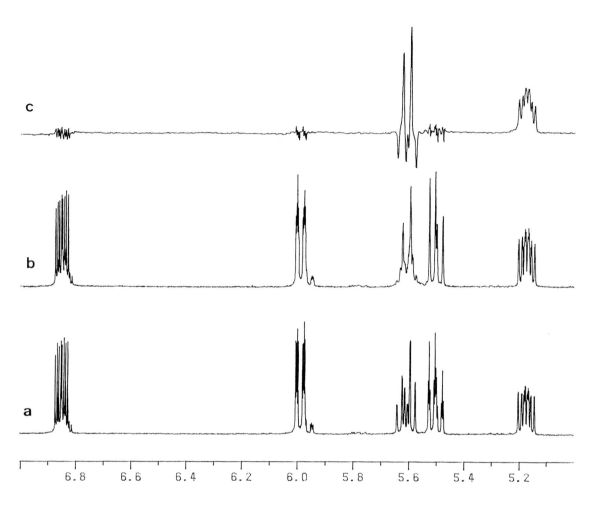

Fig. 3.2.2. (a) Section of the ¹H NMR spectrum of **2**; (b) selective ¹H decoupling experiment, irradiation at $\delta = 2.03$; (c) NOE difference experiment, irradiation at $\delta = 2.03$.

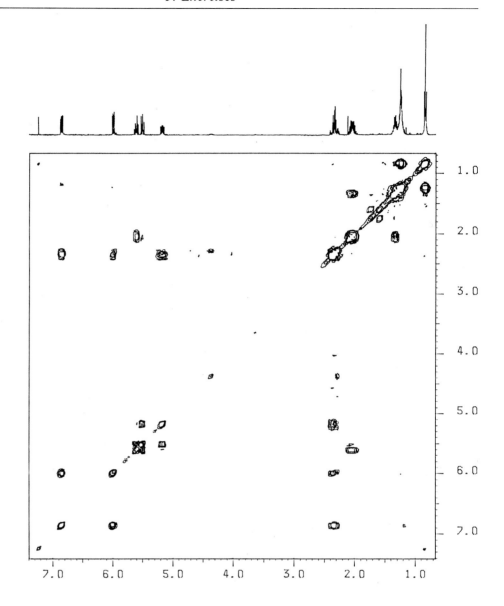

Fig. 3.2.3. H,H COSY spectrum of **2**.

Fig. 3.2.4. H,C COSY and DEPT spectra of **2**; the ^{13}C NMR spectrum contains an additional signal at $\delta = 164.5$ (C).

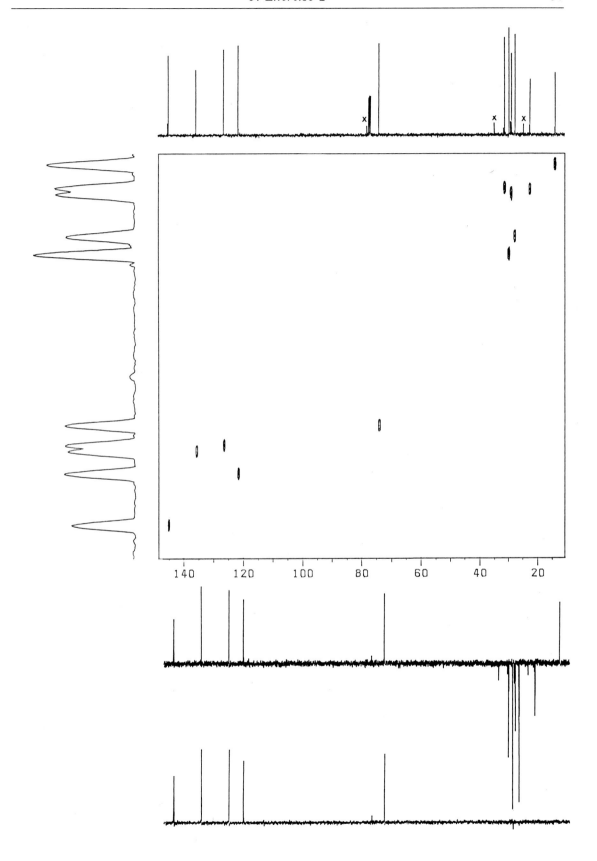

Exercise 3

Figure 3.3.1 shows the ^{1}H NMR spectrum of 4[e]-bromoadamantanone[1](3). Assign all ^{1}H and ^{13}C signals with the help of the H,H and H,C COSY spectra. Discuss the multiplicities of the proton signals.

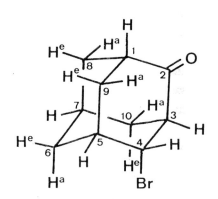

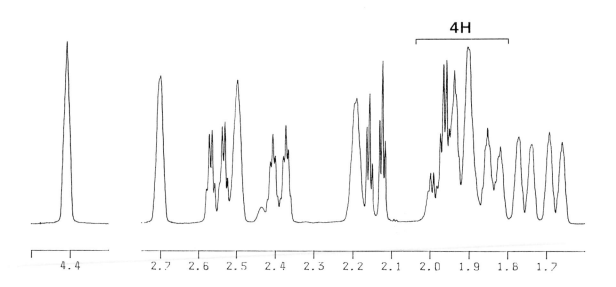

Fig. 3.3.1. ^{1}H NMR spectrum of **3**.

[1]Nomenclature: The superfixes "e" and "a" denote equatorial and axial, respectively, referring to the stereochemical position of the corresponding atom relative to the cyclohexane unit bearing the highest number of substituents.

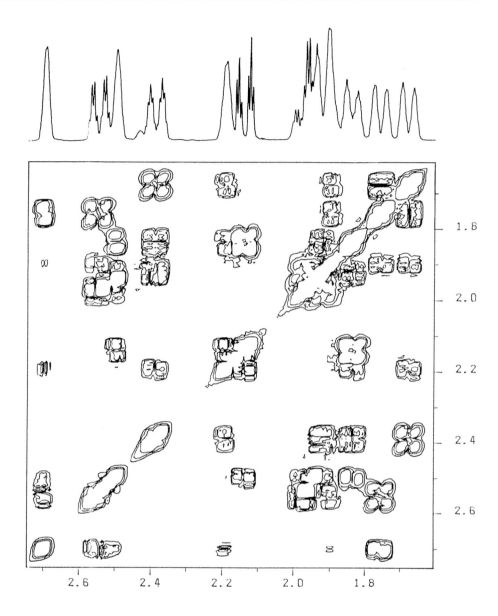

Fig. 3.3.2. Expanded section of the H,H COSY spectrum of **3**.

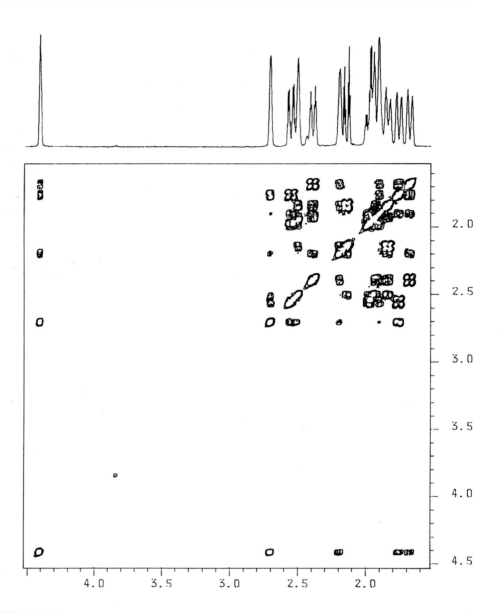

Fig. 3.3.3. H,H COSY spectrum of **3**.

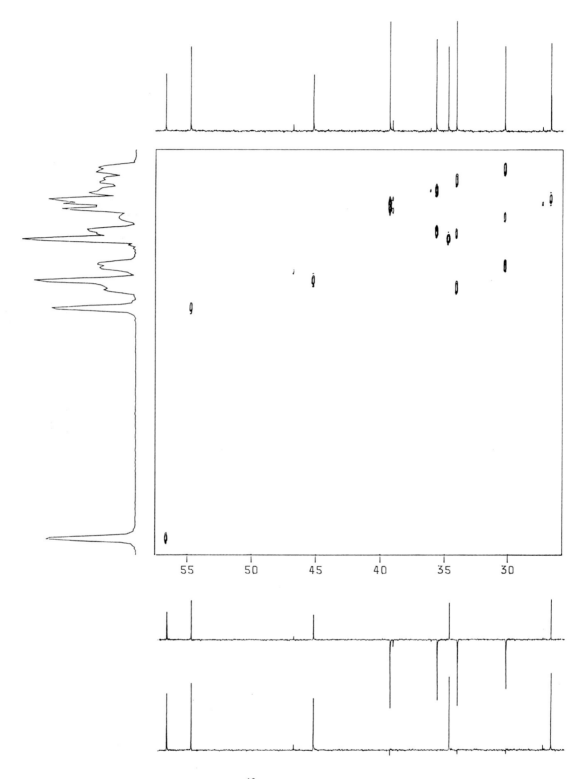

Fig. 3.3.4. H,C COSY and DEPT spectra of **3**; the ^{13}C NMR spectrum contains an additional signal at $\delta = 212.5$ (C).

Exercise 4

The hydrobromic acid catalyzed rearrangement of the bromolactone **4** resulted in the three isomeric dibromoadamantanones **5** through **7**. To which of them belong the NMR spectra depicted in Figs. 3.4.1 through 3.4.3? For the nomeclature see Exercise 3.

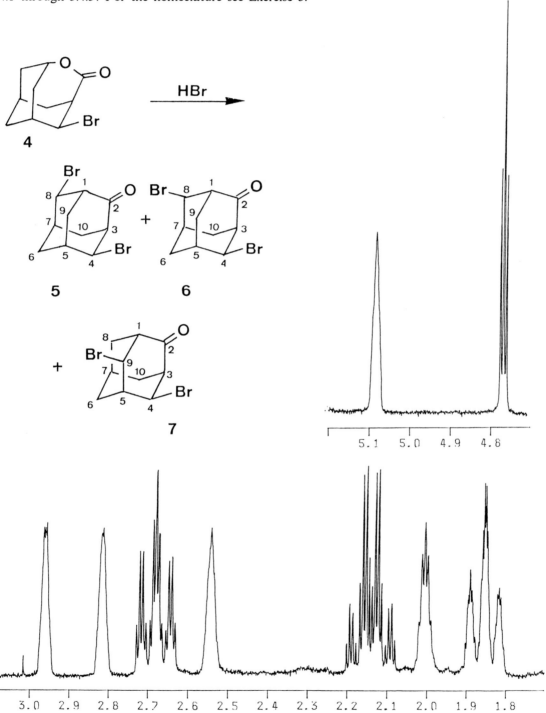

Fig. 3.4.1. ^{1}H NMR spectrum of a dibromoadamantanone.

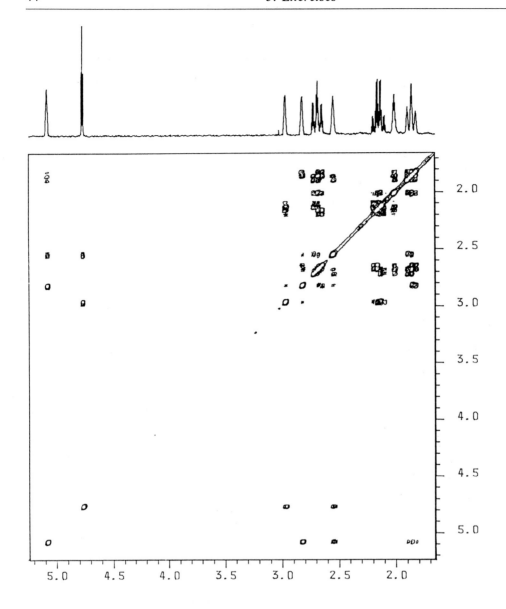

Fig. 3.4.2. H,H COSY spectrum of a dibromoadamantanone.

Fig. 3.4.3. H,C COSY and DEPT spectra of a dibromoadamantanone; the ^{13}C NMR spectrum contains an additional signal at $\delta = 208.7$ (C).

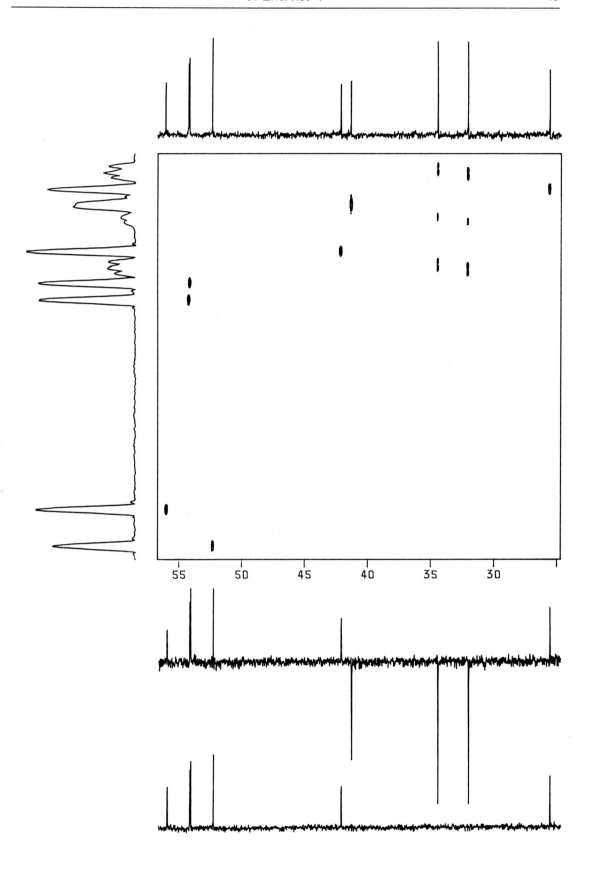

Exercise 5

The reduction of a naturally occurring carboxylic acid produced an alcohol with a molecular formula $C_{10}H_{17}OH$. What is its structure?

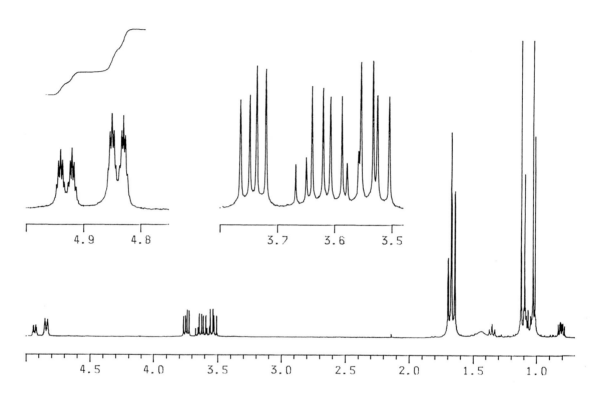

Fig. 3.5.1. ^1H NMR spectrum; the signals at $\delta = 4.93$ and 4.84 have been integrated, and the intensity ratio is about 1:2.

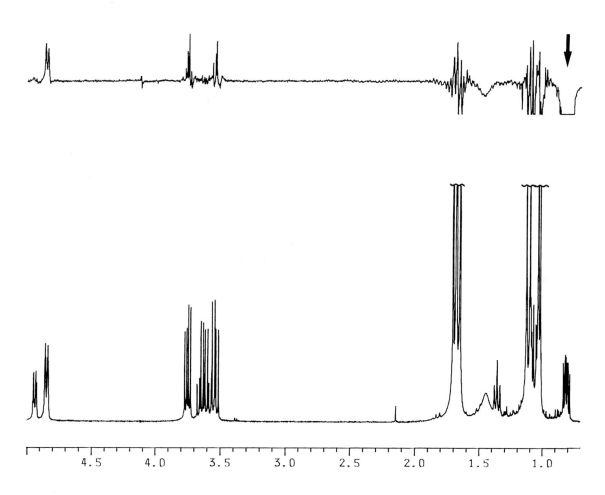

Fig. 3.5.2. NOE difference spectrum, irradiation at $\delta = 0.80$; the lower trace is the reference spectrum.

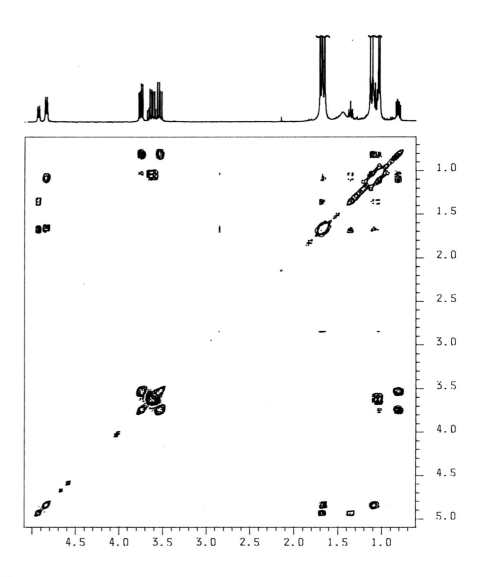

Fig. 3.5.3. H,H COSY spectrum.

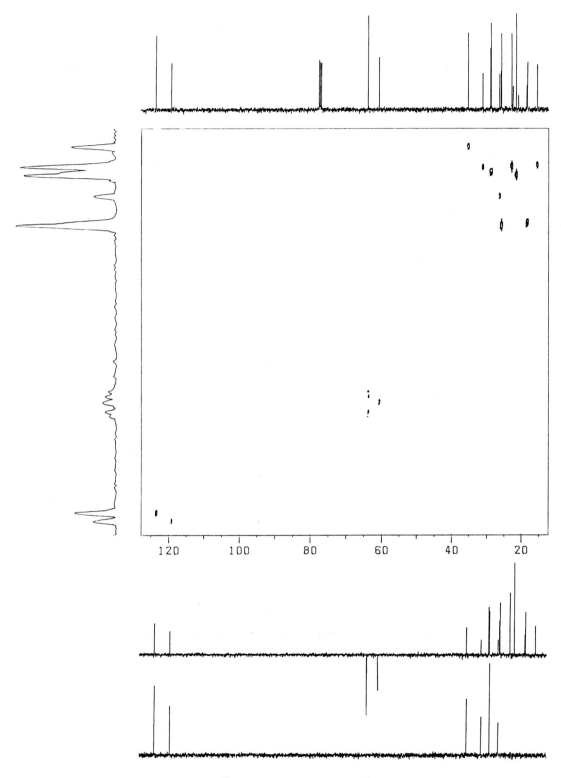

Fig. 3.5.4. H,C COSY and DEPT spectra. The ^{13}C NMR spectrum contains two additional signals at $\delta = 135.0$ (C) and $\delta = 133.0$ (C); the latter is more intensive.

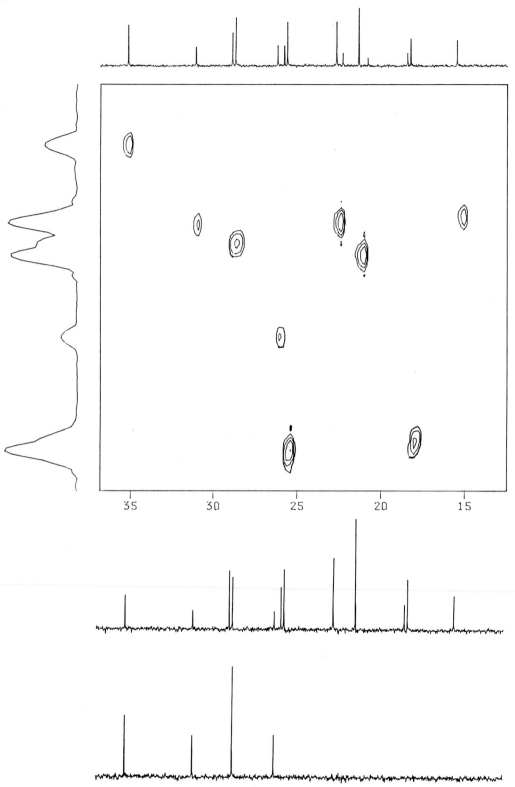

Fig. 3.5.5. Expanded sections of the H,C COSY and DEPT spectra.

Exercise 6

After the reaction of 2-aminothiophenol with a bicyclic α,β-unsaturated ketone, a thiazepin **12** was obtained. What information about the configuration and the preferred conformation of compound **12** can be extracted from the NMR spectra?

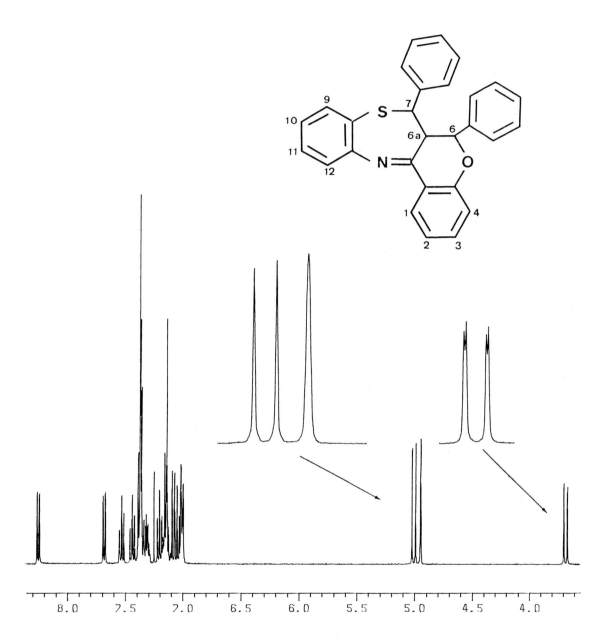

Fig. 3.6.1. ^{1}H NMR spectrum of **12**.

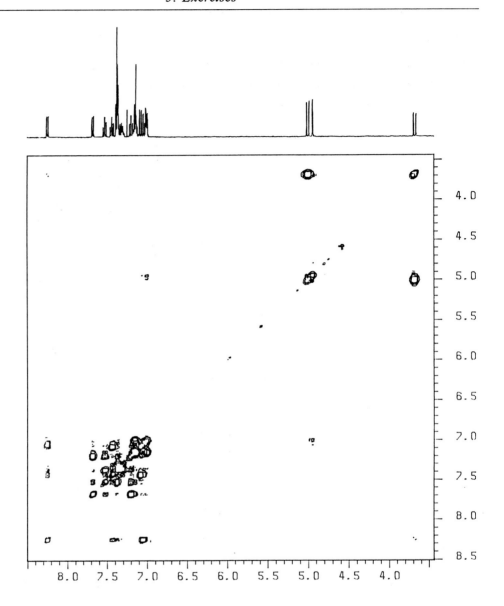

Fig. 3.6.2. H,H COSY spectrum of **12**.

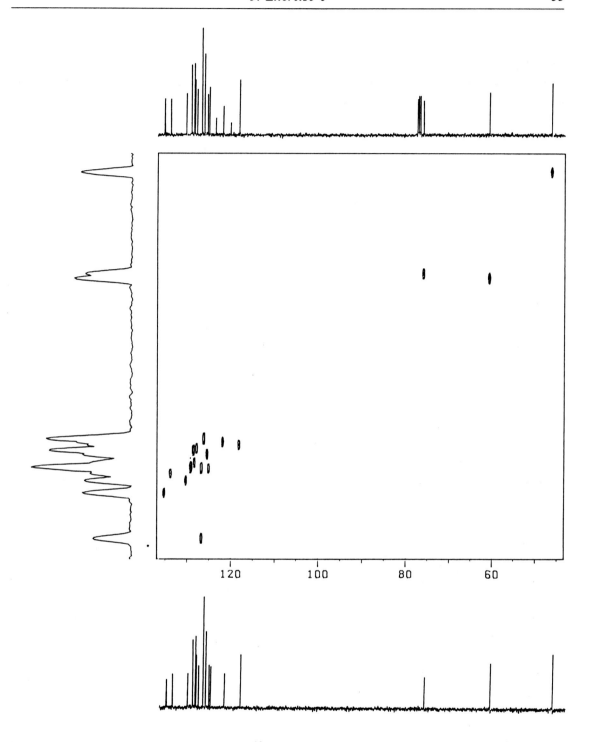

Fig. 3.6.3. H,C COSY and DEPT spectra of **12**; the ^{13}C NMR spectrum contains five additional signals at δ = 162.0, 155.5, 151.9, 142.7, and 138.1 (all C). Since there are only quarternary and methine carbons in the molecule, the second DEPT spectrum is omitted.

Exercise 7

The reaction of cyclobutylidene and an olefin with the molecular formula C_4H_6 produced a hydrocarbon **13** with an unusual structure. In the mass spectrum an M^+ peak with $m/e = 108$ was found. What is the structure of **13**? Assign all ^1H and ^{13}C signals.

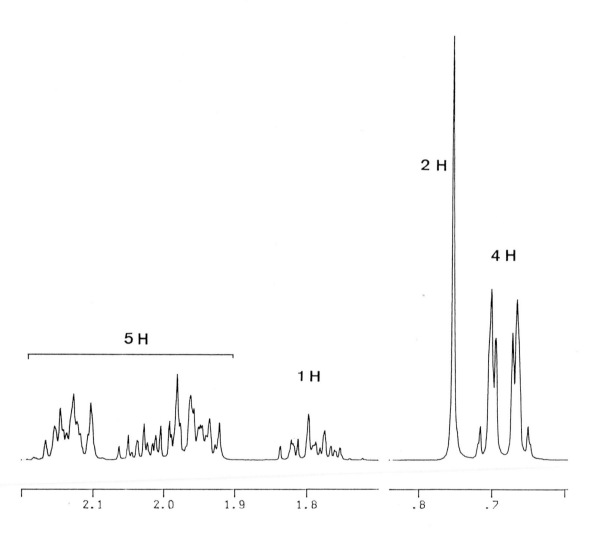

Fig. 3.7.1. ^1H NMR spectrum of **13**.

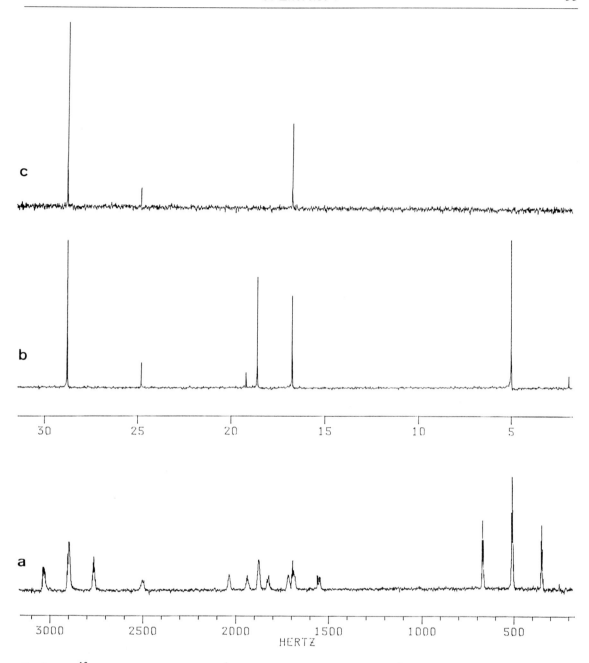

Fig. 3.7.2. ^{13}C NMR spectra of **13**; (a) ^{1}H coupled (gated decoupled); (b) ^{1}H broad-band decoupled (BB); (c) hetero-NOE experiment, the proton at $\delta = 2.17–2.10$ is irradiated.

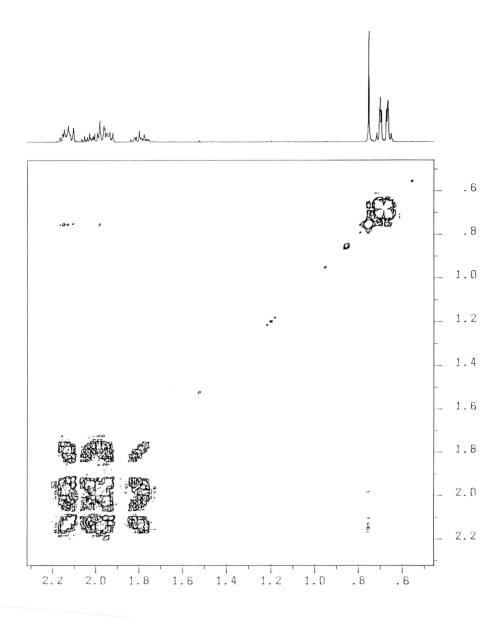

Fig. 3.7.3. H,H COSY spectrum of **13**.

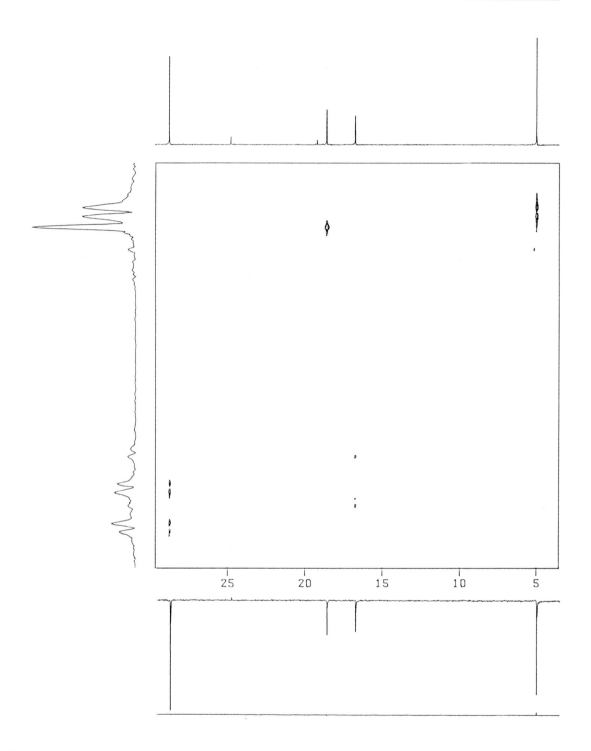

Fig. 3.7.4. H,C COSY and DEPT spectra of **13**.

Exercise 8

An oil with an ethereal smell was isolated from *Thymus hyemalis L.*, a plant growing in Spain. The compound (14) has the molecular formula $C_{10}H_{14}O$. What is its structure? Assign all ^1H and ^{13}C signals.

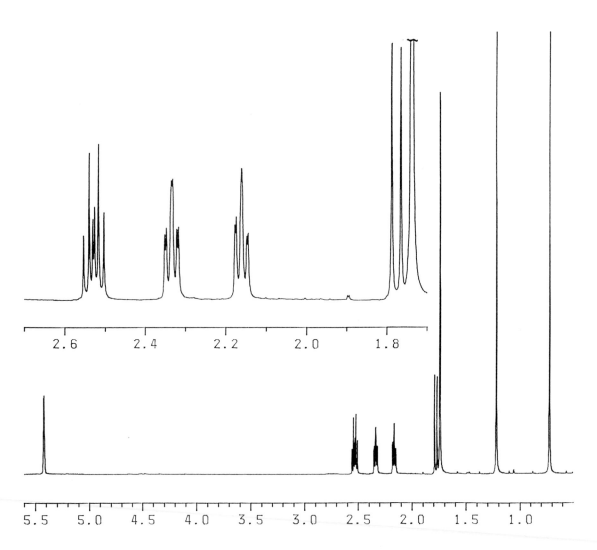

Fig. 3.8.1. ^1H NMR spectrum of 14; the signal at $\delta = 1.74$ is split into a doublet (J = 1.2 Hz), and its intensity corresponds to three hydrogens.

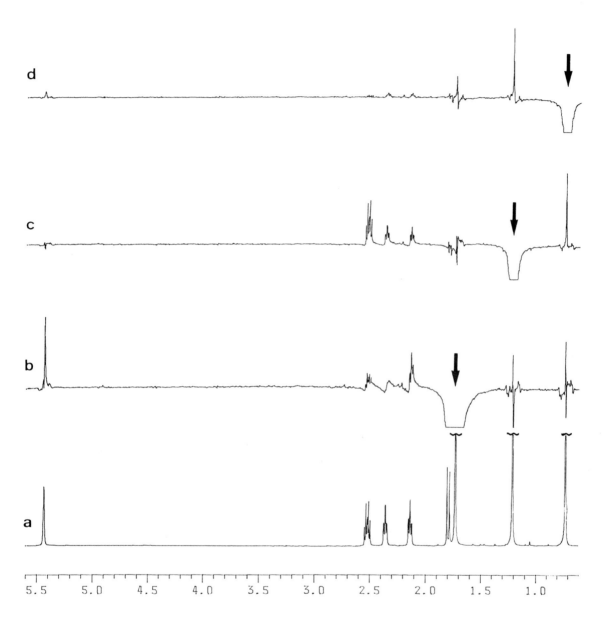

Fig. 3.8.2. NOE difference spectra of **14**: (a) Reference spectrum. Irradiations at (b) $\delta = 1.74$, (c) $\delta = 1.22$, (d) $\delta = $ 0.72.

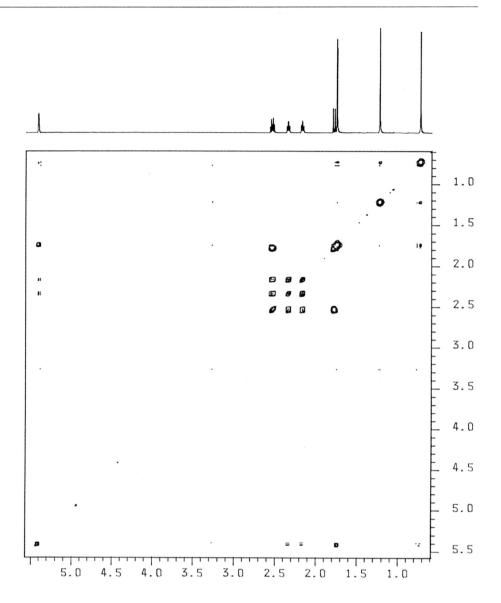

Fig. 3.8.3. H,H COSY spectrum of **14**.

Fig. 3.8.4. H,C COSY and DEPT spectra of **14**; the ^{13}C NMR spectrum contains two additional signals at $\delta = 203.3$ (C) and 169.9 (C).

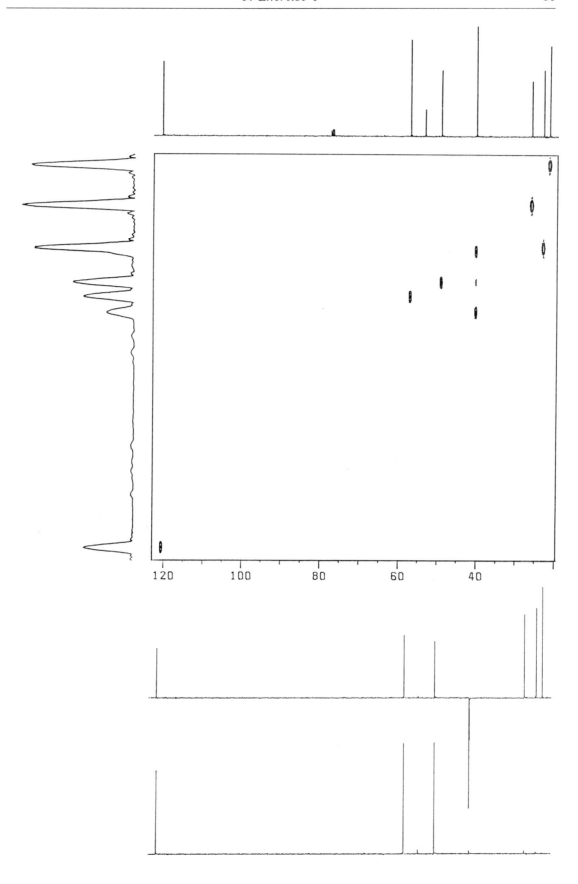

Exercise 9

In a multistep steroid synthesis an intermediate compound with formula **15** was obtained. Determine the configuration at C–16 and C–20 from the NMR spectra. The stereochemistry at C–14 and C–17 is known from the precursor.

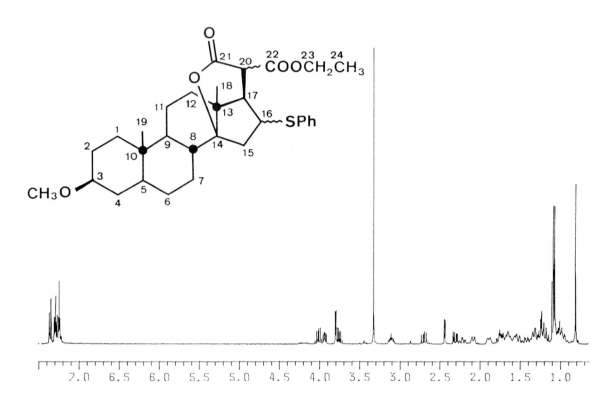

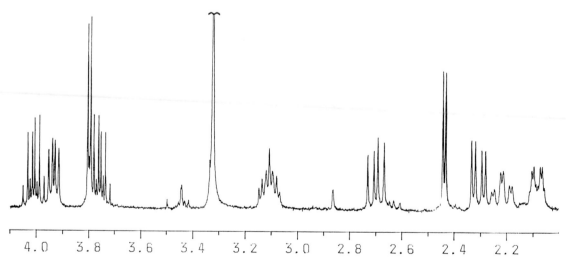

Fig. **3.9.1.** ^{1}H NMR spectrum of **15**.

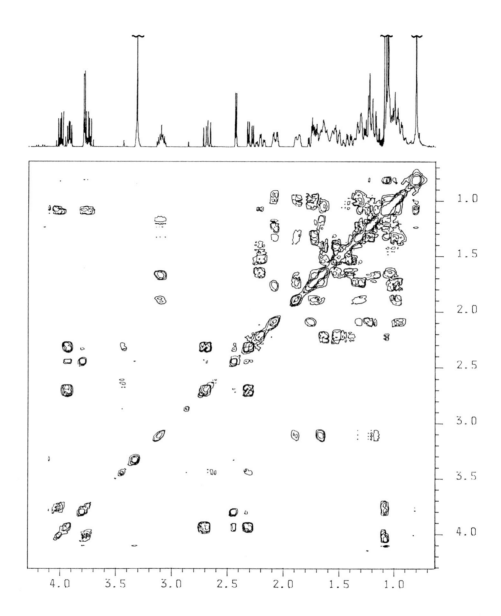

Fig. 3.9.2. Expanded section of the H,H COSY spectrum of **15**.

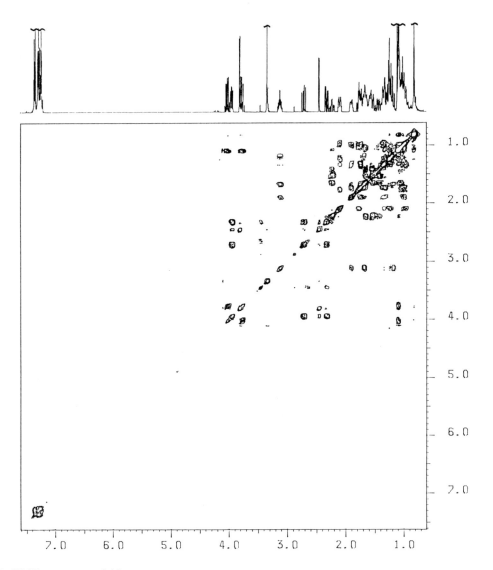

Fig. 3.9.3. H,H COSY spectrum of **15**.

Fig. 3.9.4. H,C COSY and DEPT spectra of **15**; the ^{13}C NMR spectrum contains three additional signals at $\delta = 168.5$ (C), 166.2 (C), and 135.6 (C).

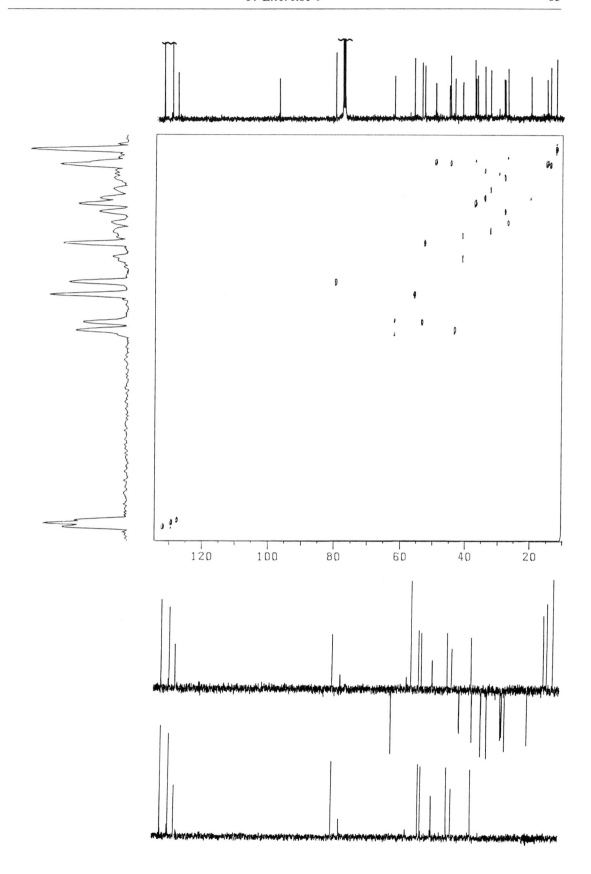

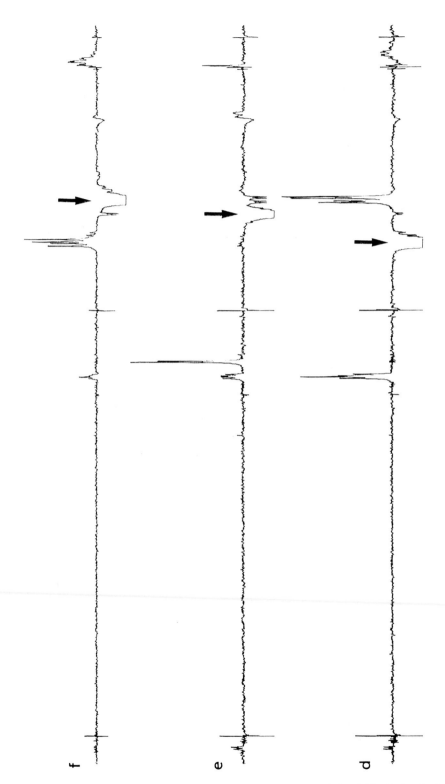

Fig. 3.9.5. NOE difference spectra of **15**: (a) Reference spectrum. Irradiation at (b) $\delta = 3.93$, (c) $\delta = 3.79$, (d) $\delta = 2.70$, (e) $\delta = 2.44$, (f) $\delta = 2.31$.

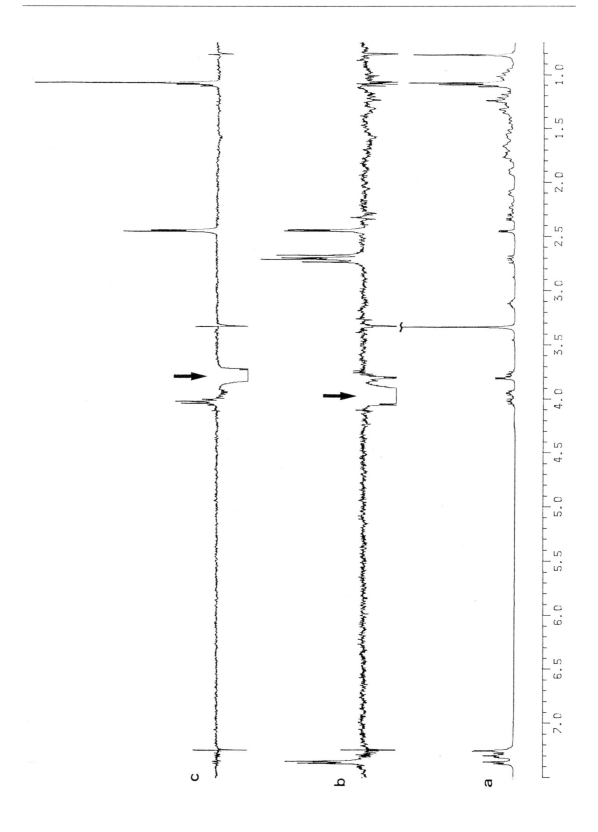

Exercise 10

A sequiterpene lactone **16** has been isolated from the Ethiopian plant *Artemisia rehan*. Two constitution formulas, **17** and **18**, have been proposed:
Which is the correct one and what is the relative configuration of this compound? Assign all signals.

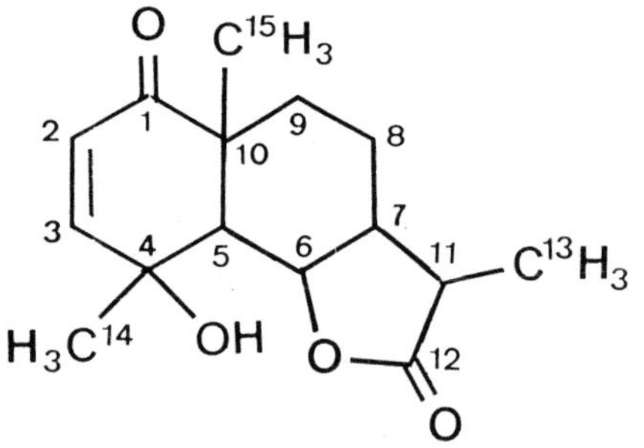

17 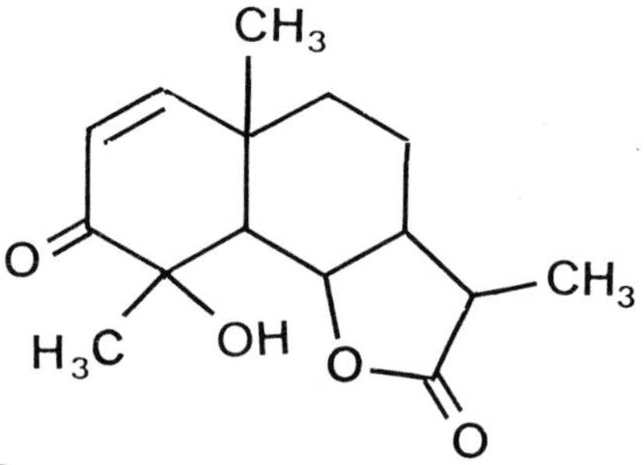   **18**

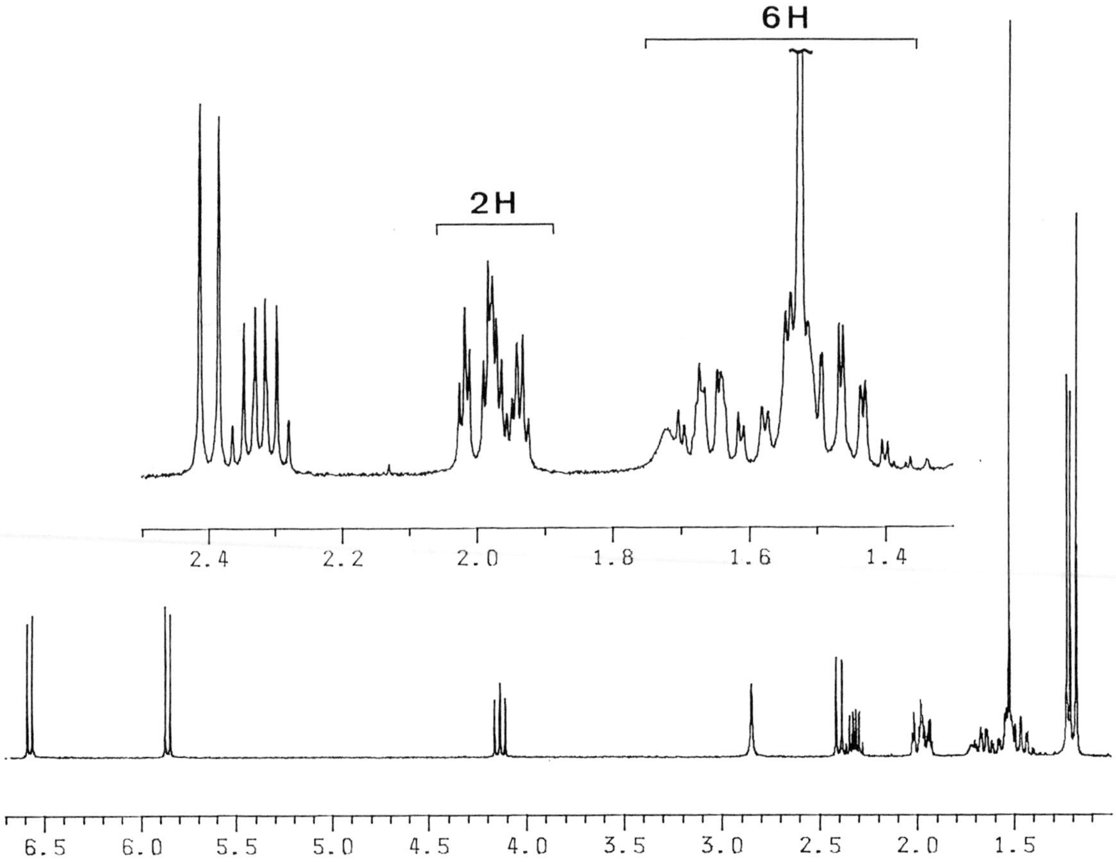

Fig. 3.10.1. ^{1}H NMR spectrum of **16**.

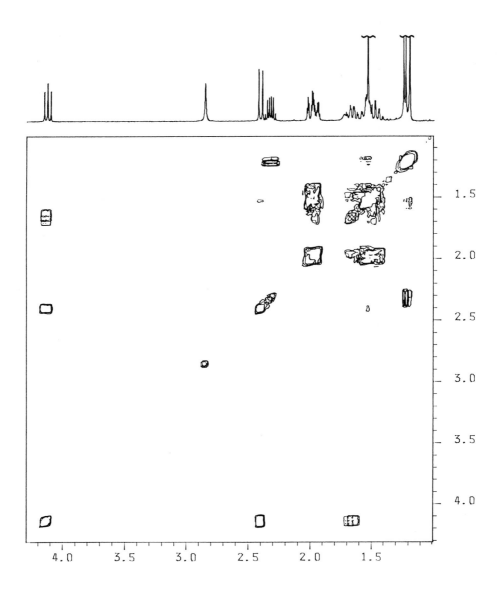

Fig. 3.10.2. Expanded section of the H,H COSY spectrum of **16**.

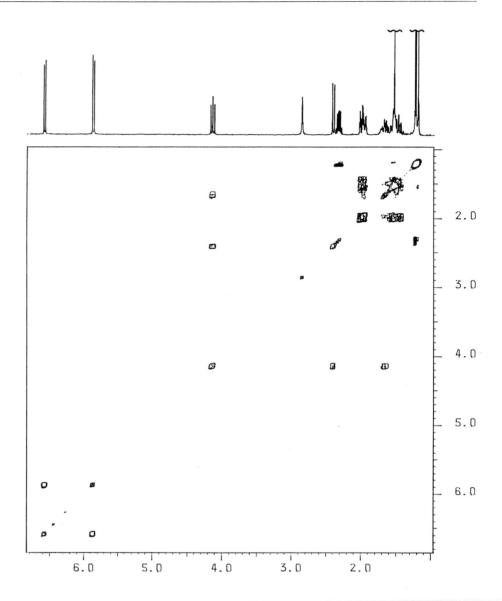

Fig. 3.10.3. H,H COSY spectrum of **16**.

Fig. 3.10.4. H,C COSY and DEPT spectra of **16**; the ^{13}C NMR spectrum contains two additional signals at $\delta = 201.6$ (C) and 178.2 (C).

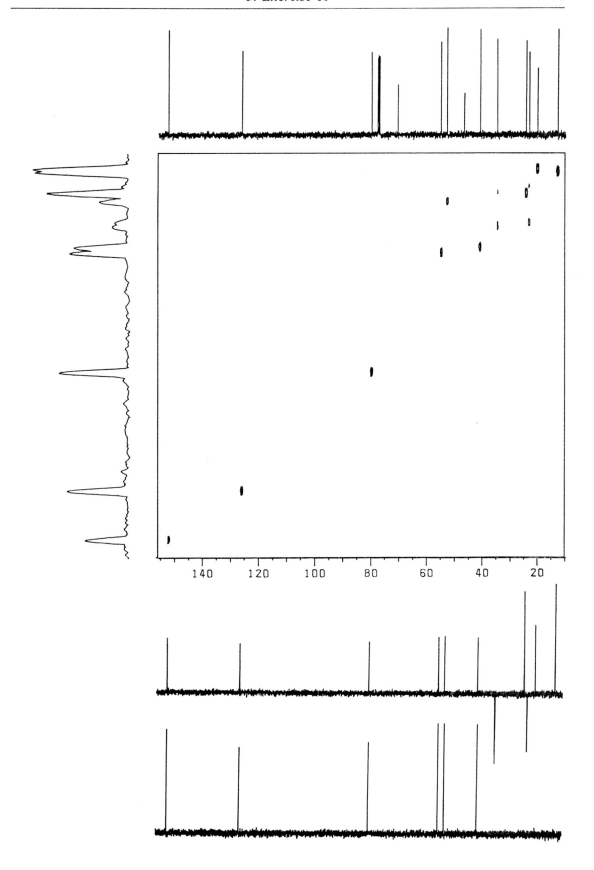

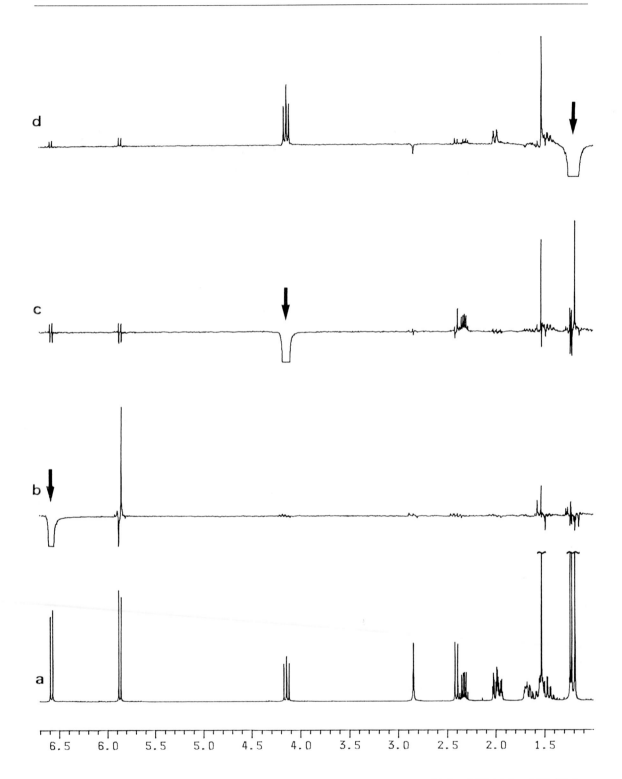

Fig. 3.10.5. NOE difference spectra of **16**: (a) Reference spectrum. Irradiation at (b) $\delta = 6.57$, (c) $\delta = 4.13$, (d) $\delta = 1.18$.

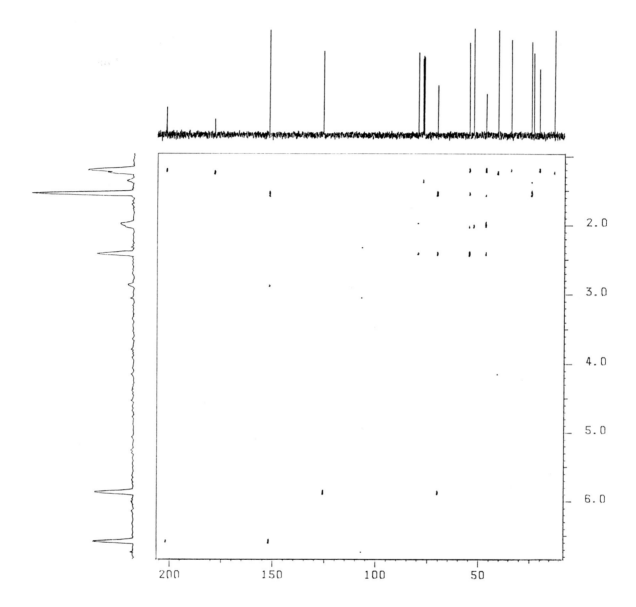

Fig. 3.10.6. COLOC spectrum of **16**.

Exercise 11

An amino acid (**19**) occurring in the red algae *Diginea simplex* has the molecular formula $C_{10}H_{15}NO_4$. Which amino acid is it? The substance was dissolved in deuterated dimethylsulfoxide and recorded at 60° C.

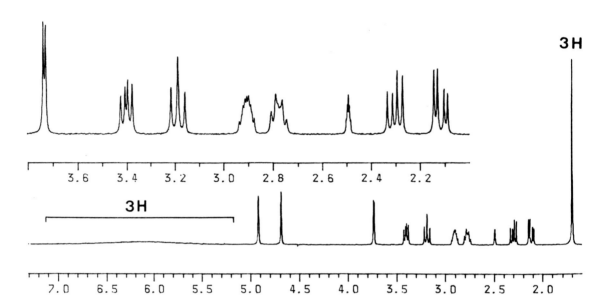

Fig. 3.11.1. ^{1}H NMR spectrum of **19**, in DMSO–d_6.

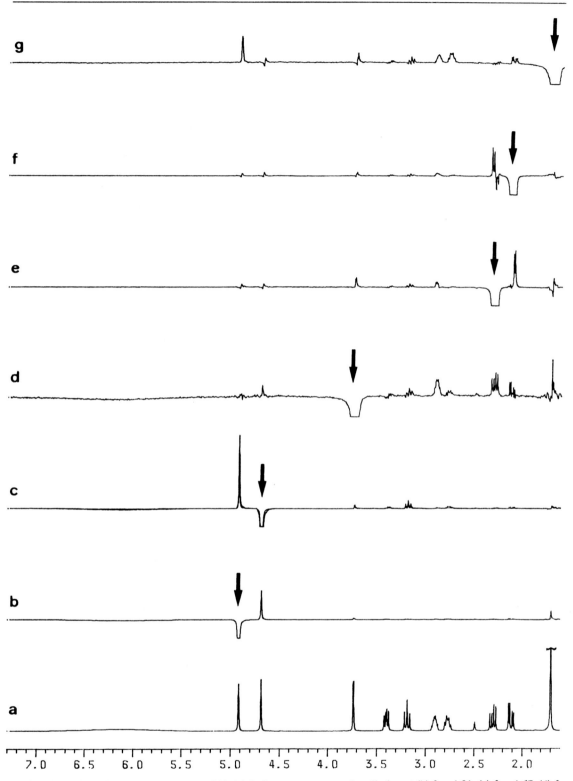

Fig. 3.11.2. NOE difference experiment of **19**: (a) Reference spectrum. Irradiation at (b) δ = 4.91, (c) δ = 4.67, (d) δ = 3.73, (e) δ = 2.29, (f) δ = 2.12, (g) δ = 1.70.

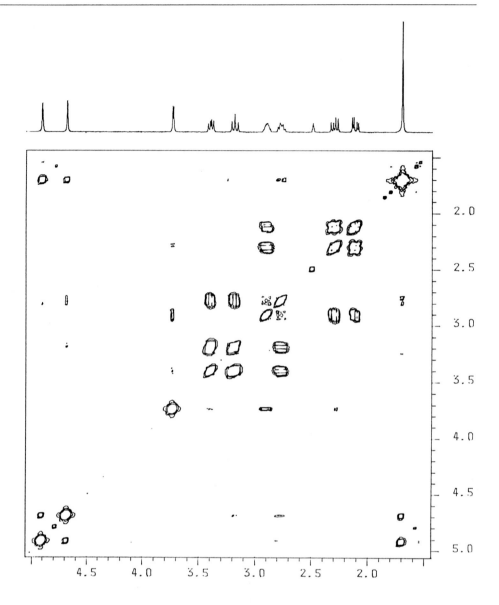

Fig. 3.11.3.   H,H COSY spectrum of **19**, in DMSO–d_6.

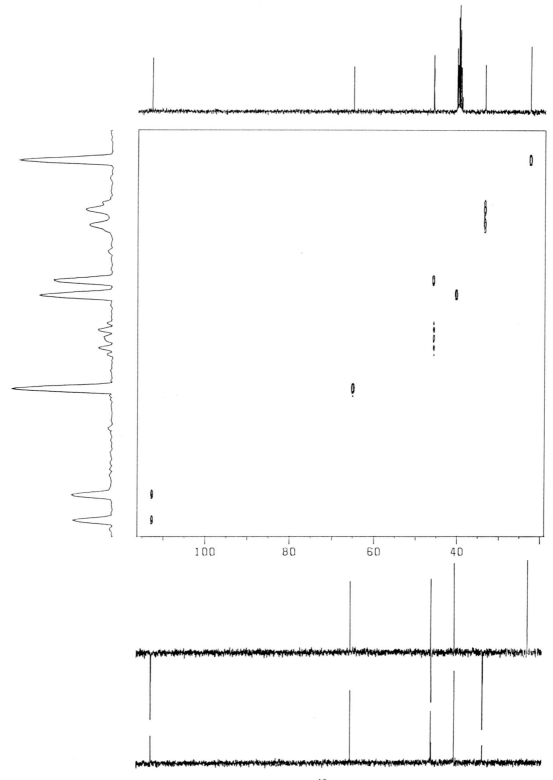

Fig. 3.11.4. H,C COSY and DEPT spectra of **19** in DMSO–d_6; the ^{13}C NMR spectrum contains three additional signals at $\delta = 172.7$ (C), 169.3 (C), and 140.6 (C).

Exercise 12

A highly unsaturated compound **20** with the molecular formula $C_{19}H_{22}O_3$ was isolated from the Sudanese plant *Peucedanum grantii Kingston*. Its IR spectrum shows a band at 1725 cm^{-1} and a very broad band between 3400 and 3600 cm^{-1}. What is its structure?

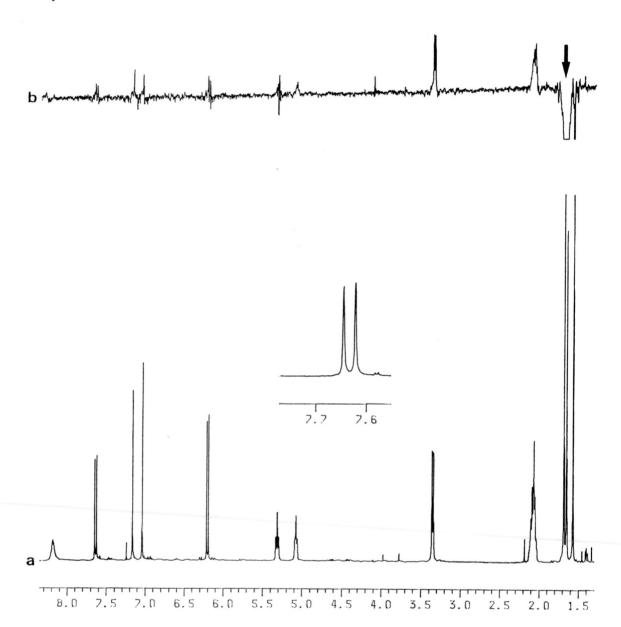

Fig. 3.12.1. (a) ^{1}H NMR of **20**: The signal at $\delta = 8.16$ can be removed by adding D_2O and shaking the chloroform solution. (b) NOE difference spectrum, irradiation at $\delta = 1.69$.

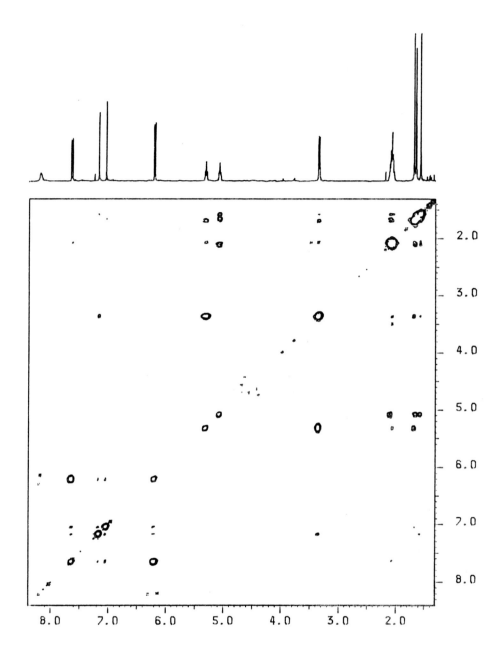

Fig. 3.12.2. H,H COSY spectrum of **20**.

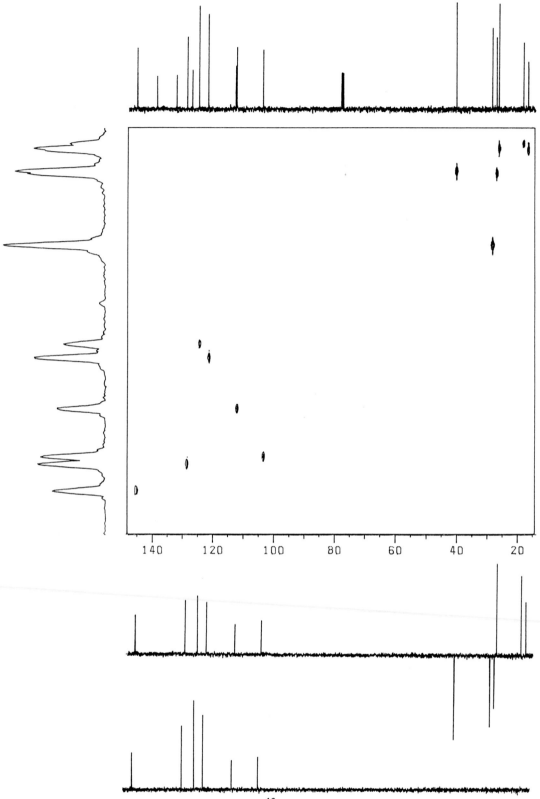

Fig. 3.12.3. H,C COSY and DEPT spectra of **20** ; the ^{13}C NMR spectrum contains three additional signals at δ = 162.9, 158.8, and 153.9.

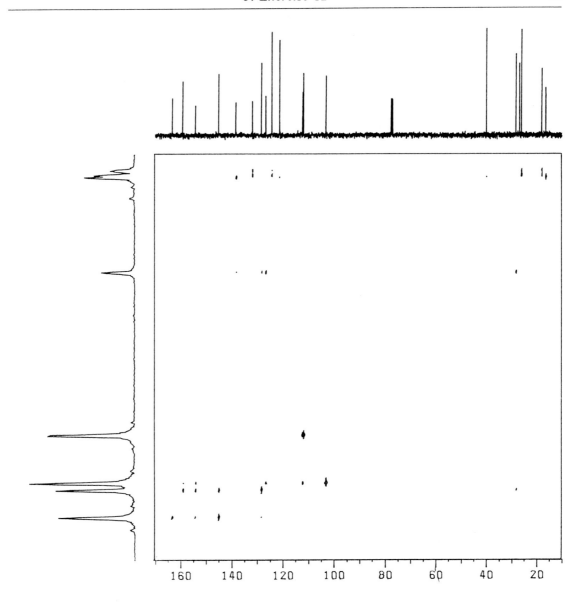

Fig. 3.12.4. COLOC spectrum of **20**.

Exercise 13

The glycoside **21** has the molecular formula $C_{18}H_{20}N_2O_{12}$. What is the structure of the sugar part and of the aglycone? Is **21** an α- or β-glycoside?

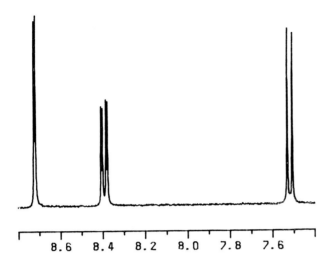

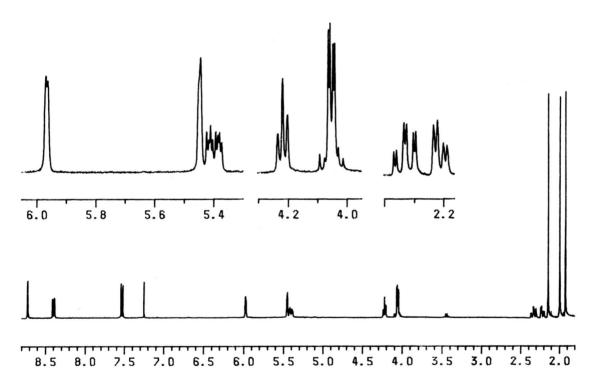

Fig. 3.13.1. ^{1}H NMR spectrum of **21**.

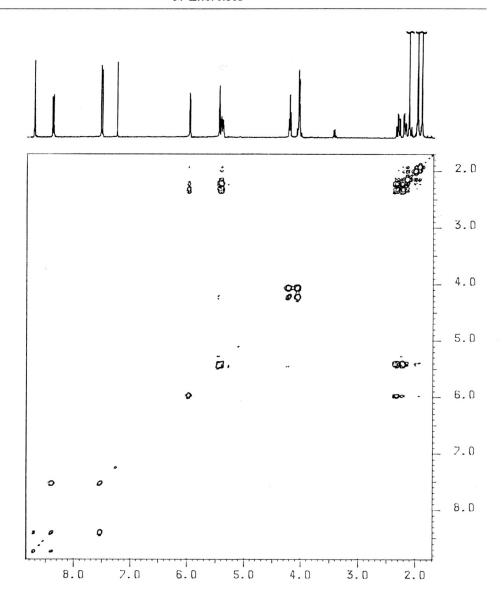

Fig. 3.13.2.   H,H COSY spectrum of **21**.

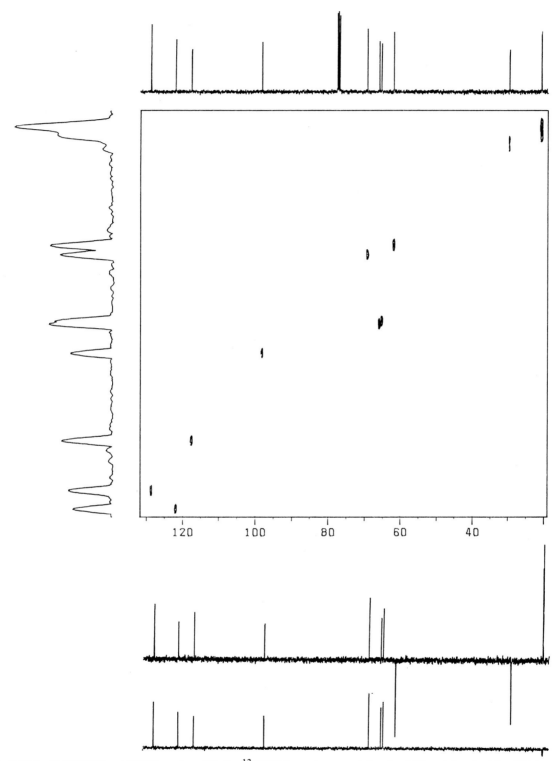

Fig. 3.13.3. H,C COSY and DEPT spectra of **21**; the ^{13}C NMR spectrum contains six additional signals at $\delta = 170.1$, 169.9, 169.3, 153.8, 141.3, and 139.7 (all C).

Exercise 14

A tetracyclic compound **22** has the constitution depicted in Fig. 3.14.1. Try to assign all NMR signals and decide what the configuration of **22** is. Is a preferred conformation indicated?

Note: All five-membered ring junctions are cis.

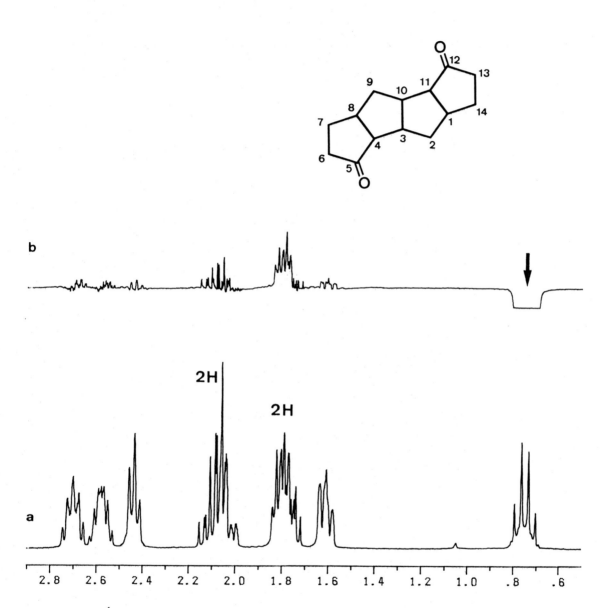

Fig. 3.14.1. (a) ^{1}H NMR spectrum of **22**; (b) NOE difference spectrum, irradiation at $\delta = 0.75$.

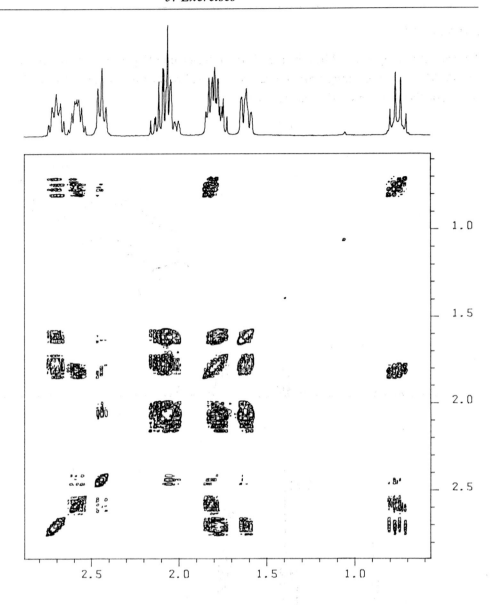

Fig. 3.14.2.   H,H COSY spectrum of **22**.

Fig. 3.14.3.   H,C COSY and DEPT spectra of **22**; the ^{13}C NMR spectrum contains an additional signal at $\delta = 221.3$.

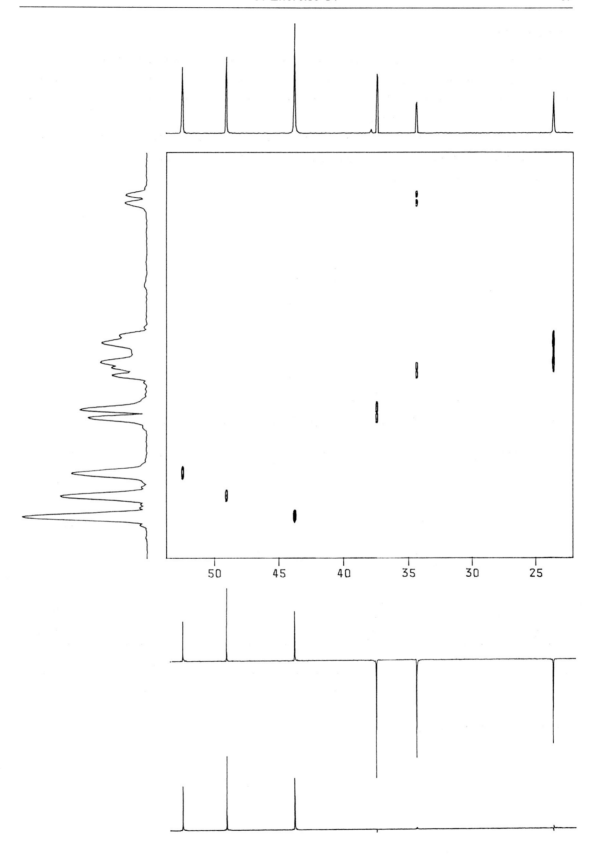

Exercise 15

A compound **23** with the molecular formula $C_{18}H_{16}O_5$ has been isolated from the Ethiopian plant *Taverniera abyssinica*. It contains an acetate and a methoxy group that are in different aromatic rings. Determine the structure of **23**, assign all proton signals, and discuss their splittings. Where is the acetate and where is the methoxy group?

In the structural elucidation of natural products it frequently happens that only a few milligrams of the substance are available, as in the present case. Therefore, even with 15 hours of spectrometer time a satisfactory H,C COSY could not be obtained.

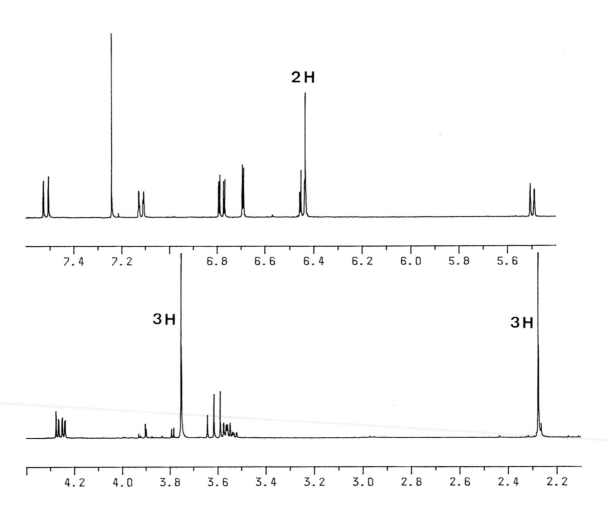

Fig. 3.15.1. ^{1}H NMR spectrum of **23**.

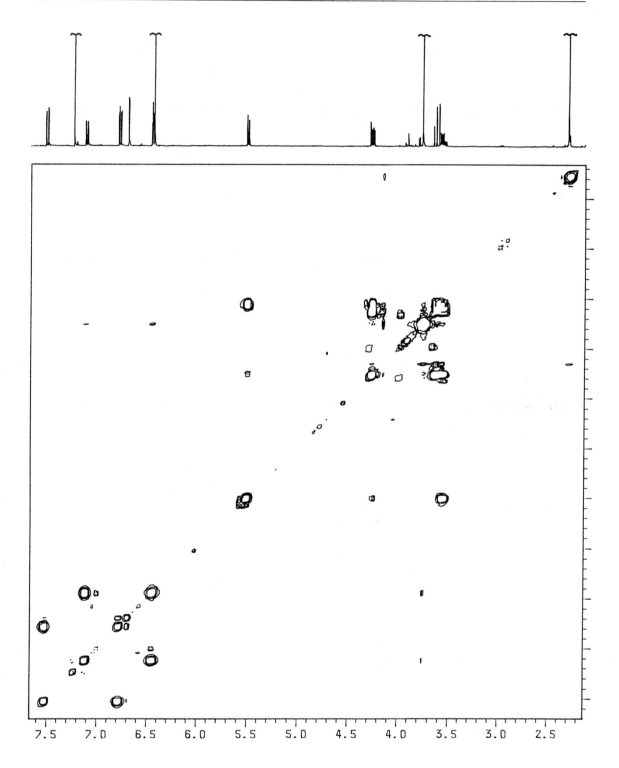

Fig. 3.15.2. H,H COSY spectrum of **23**.

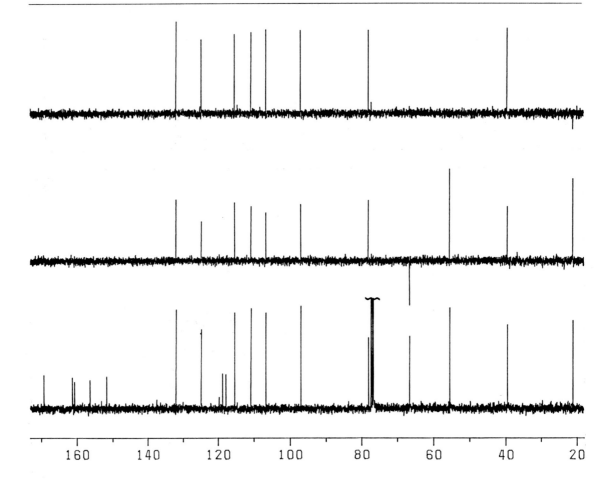

Fig. 3.15.3. ^{13}C NMR and DEPT spectra of **23**.

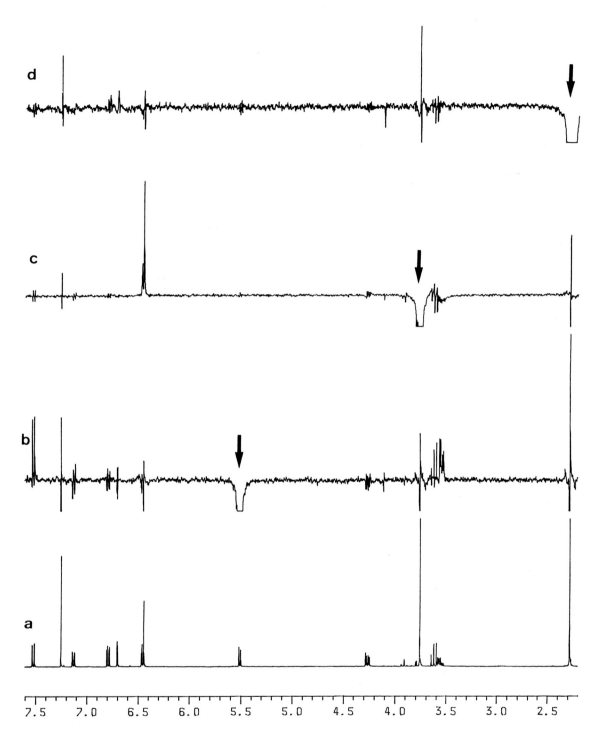

Fig. 3.15.4. NOE difference experiment of **23**: (a) Reference spectrum. Irradiation positions at (b) $\delta = 5.50$; (c) $\delta = 3.76$; (d) $\delta = 2.28$.

Exercise 16

In the literature contradictory data exist for the ^{13}C chemical shift assignments of dicyclopentadien (24)[1]. In order to make a final determination, in addition to the two COSY experiments, a 2D INA-DEQUATE experiment was performed. Assign the ^{13}C signals and compare your results with the data from the literature.

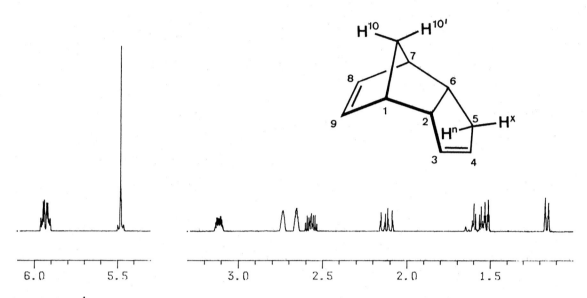

Fig. 3.16.1. ^1H NMR spectrum of **24**, in benzene–d_6.

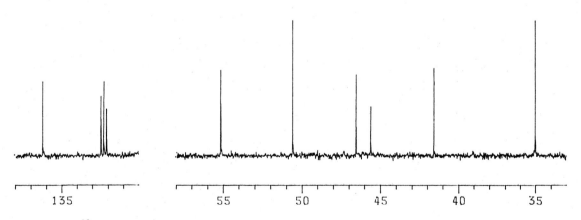

Fig. 3.16.2. ^{13}C NMR spectrum of **24**, in benzene–d_6.

[1] (a) Johnson LF, Jankowski WC (1972) Carbon-13 NMR Spectra. Wiley, New York; (b) Sadtler Standard Carbon-13 NMR Spectra (1977). Sadtler Research Laboratories, Philadelphia; (c) Nakagawa K, Iwase S, Ishi S, Hamanaka S, Okawa M (1977) *Bull Chem Soc Jpn* **50**: 2391.

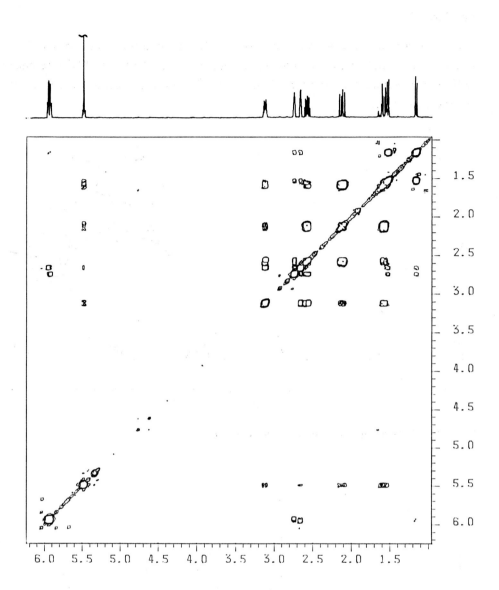

Fig. 3.16.3. H,H COSY spectrum of **24**, in benzene–d$_6$.

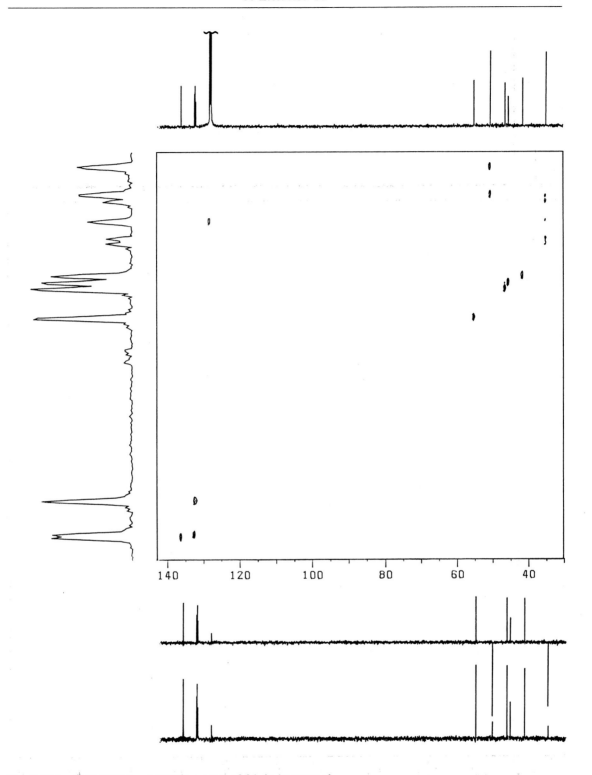

Fig. 3.16.4. H,C COSY and DEPT spectra of **24**, in benzene–d$_6$.

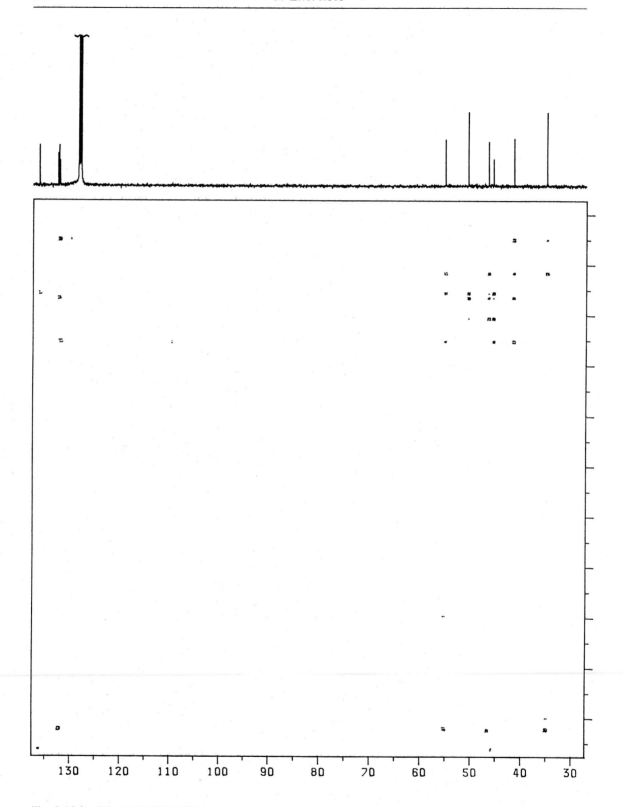

Fig. 3.16.5. 2D INADEQUATE spectrum of **24**, in benzene–d$_6$.

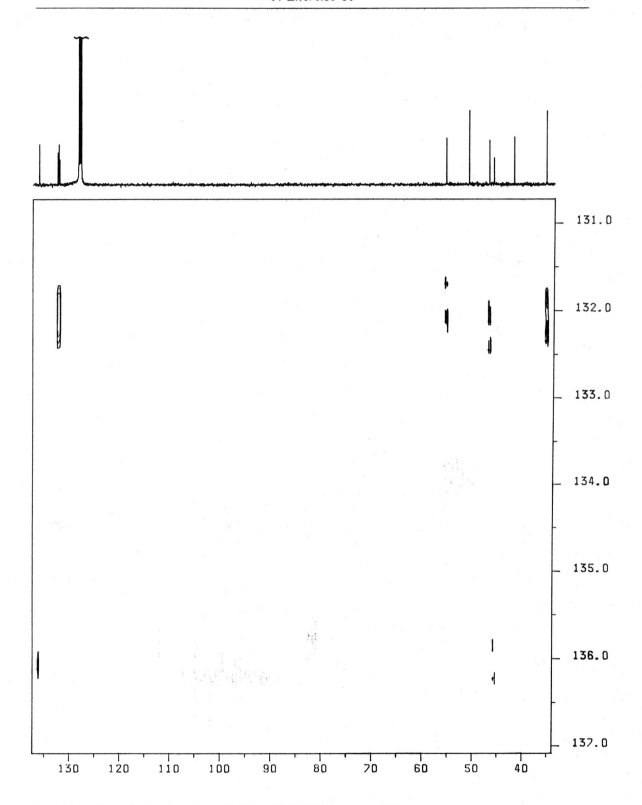

Fig. 3.16.6. Expanded section of the 2D INADEQUATE spectrum of **24.**

Exercise 17

The NMR spectra are recorded using a solution of a peracetylated methyl glycoside **25**. How many and which monosaccharide subunits does the molecule contain? How are these interconnected?

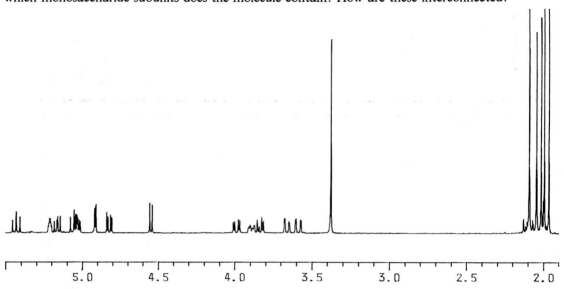

Fig. 3.17.1. ^{1}H NMR spectrum of **25**.

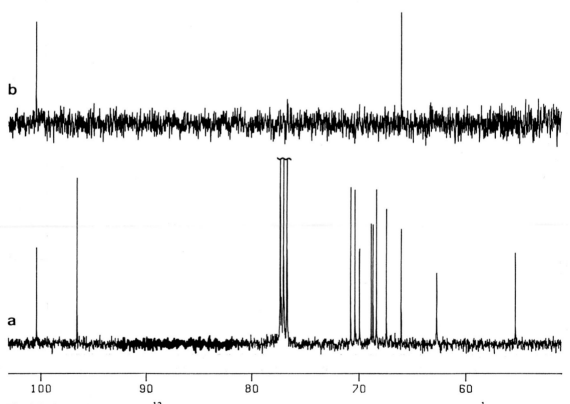

Fig. 3.17.2.   (a) Section of the ^{13}C NMR spectrum. (b) Selective INEPT experiments; pulse on the ^{1}H signal at $\delta =$ 4.55.

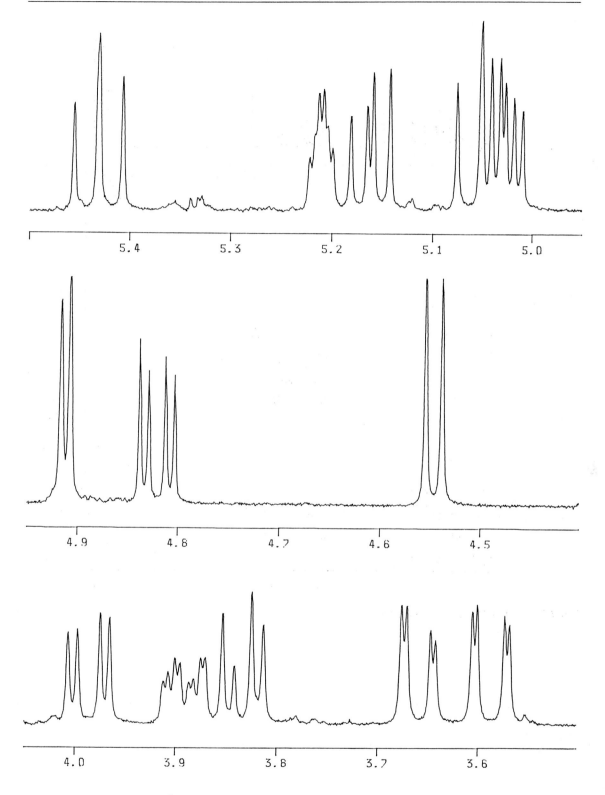

Fig. 3.17.3. Sections of the ^{1}H NMR spectrum of **25**.

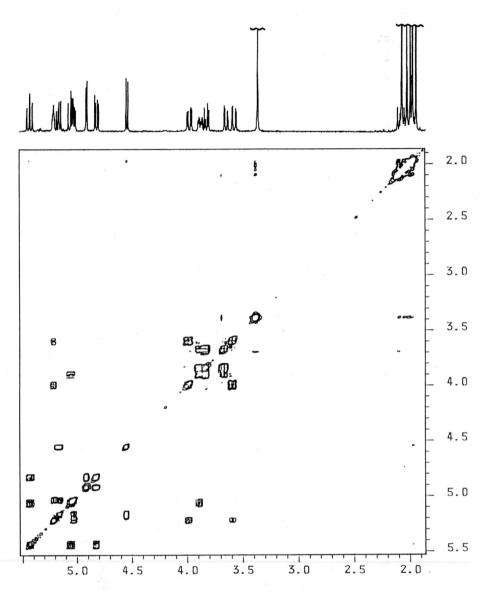

Fig. 3.17.4.   H,H COSY spectrum of **25**.

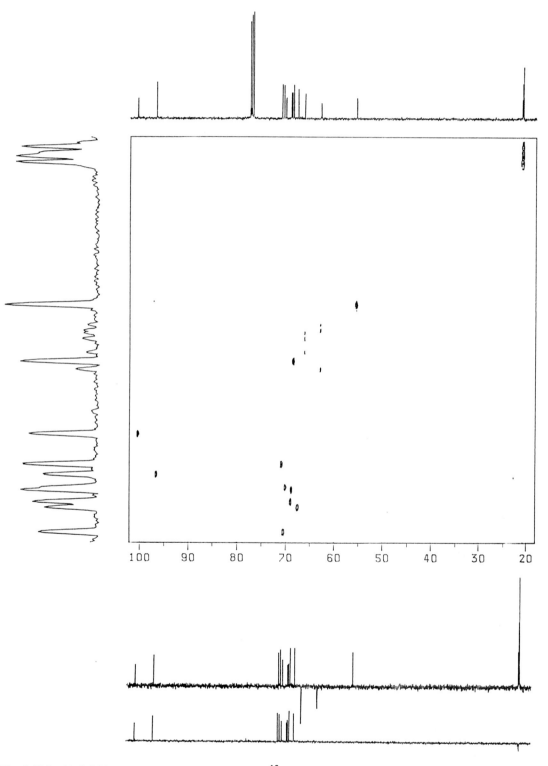

Fig. 3.17.5. H,C COSY and DEPT spectra of **25**. The ^{13}C NMR spectrum contains four additional signals at $\delta =$ 170.3 (C), 170.1 (2C), 169.4 (2C), and 169.4(C); the latter two signals are not isochronous but are very close together.

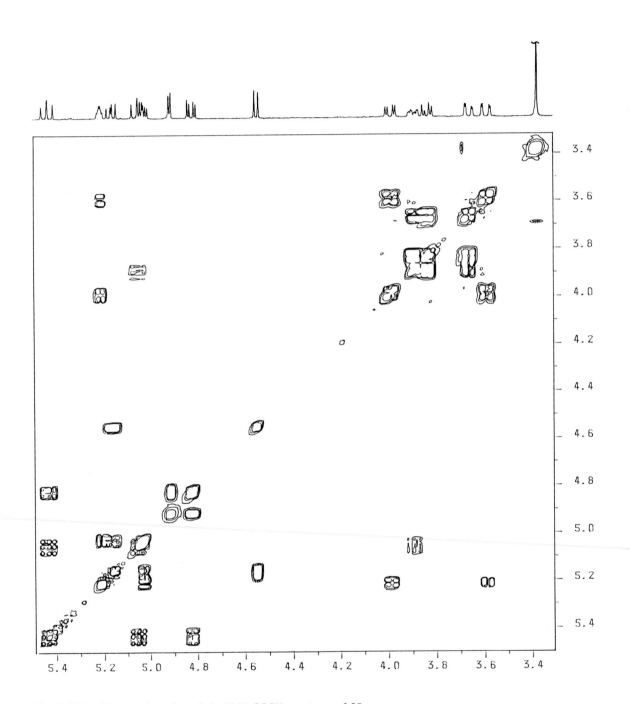

Fig. 3.17.6. Expanded section of the H,H COSY spectrum of **25**.

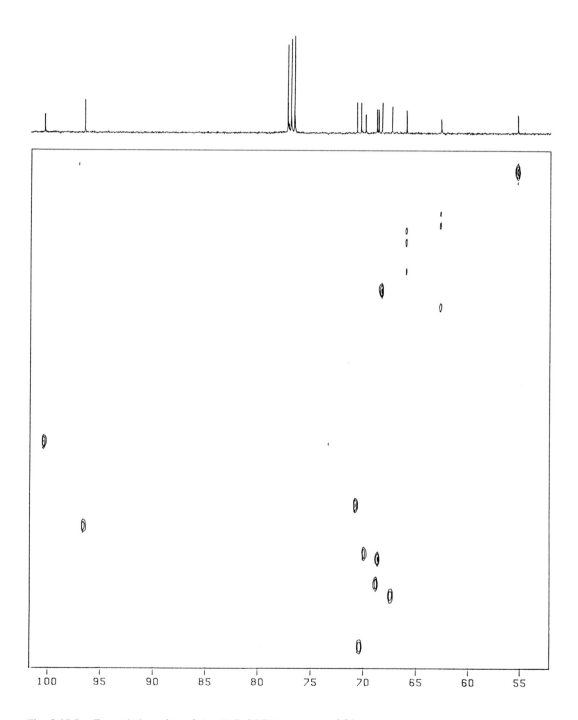

Fig. 3.17.7. Expanded section of the H,C COSY spectrum of **25**.

Exercise 18

An alkaloid **26** was isolated from the Ethiopian plant *Crotolaria rosenii*. It is a macrocyclic lactone derived from a bicyclic trialcohol **27** and dicarboxylic acid **28**.

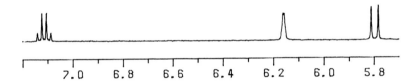

How are these structural elements interconnected and what are the relative configurations? Is there any information about the conformational behavior of compound **26**?

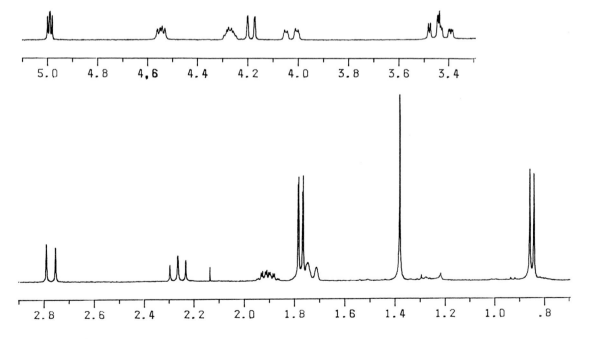

Fig. 3.18.1. ^{1}H NMR spectrum of **26**.

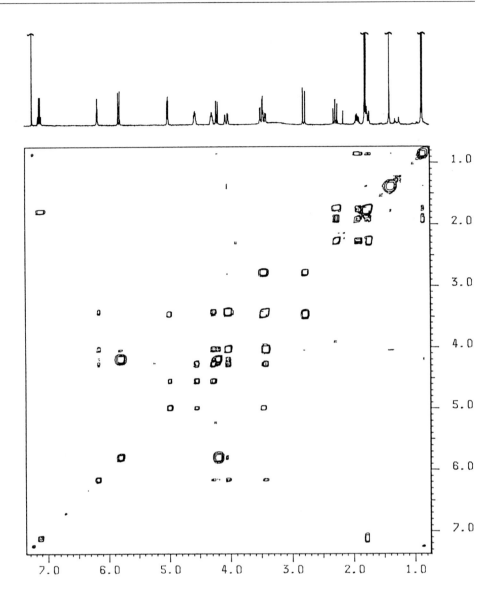

Fig. 3.18.2. H,H COSY spectrum of **26**.

Fig. 3.18.3. H,C COSY and DEPT spectra of **26**; the ^{13}C NMR spectrum contains two additional signals at $\delta = 177.0$ (C) and 167.0 (C).

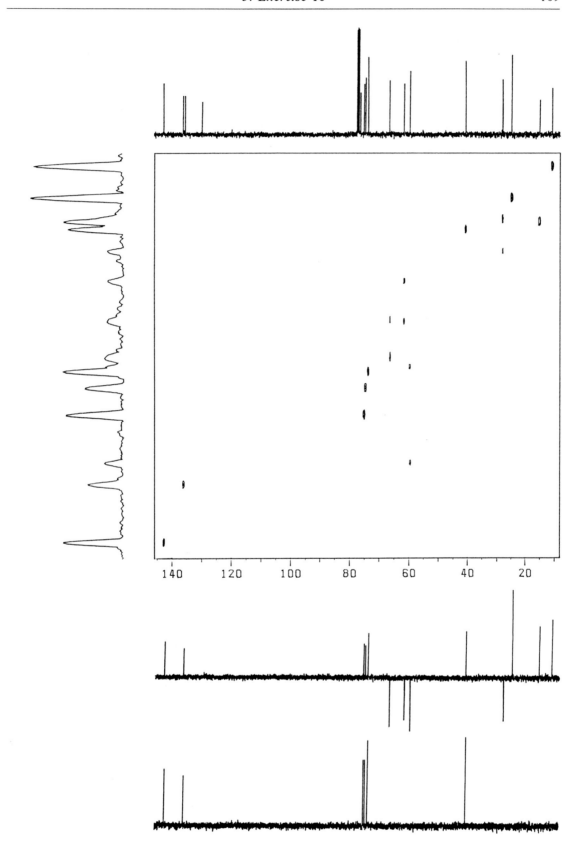

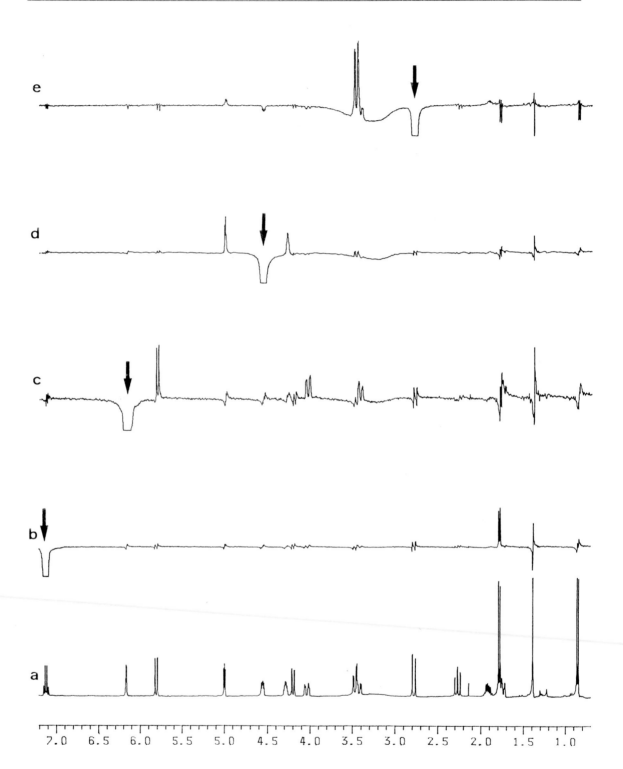

Fig. 3.18.4. NOE difference spectra of **26**: (a) Reference spectrum. Irradiation positions at (b) $\delta = 7.11$, (c) $\delta = 6.16$, (d) $\delta = 4.53$, (e) $\delta = 2.77$, (f) $\delta = 2.26$, (g) $\delta = 1.90$, (h) $\delta = 1.77$, (i) $\delta = 1.38$, (k) $\delta = 0.85$.

k

i

h

g

f

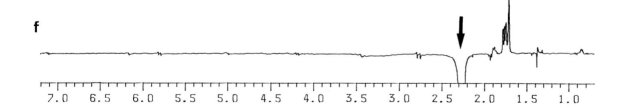

7.0   6.5   6.0   5.5   5.0   4.5   4.0   3.5   3.0   2.5   2.0   1.5   1.0

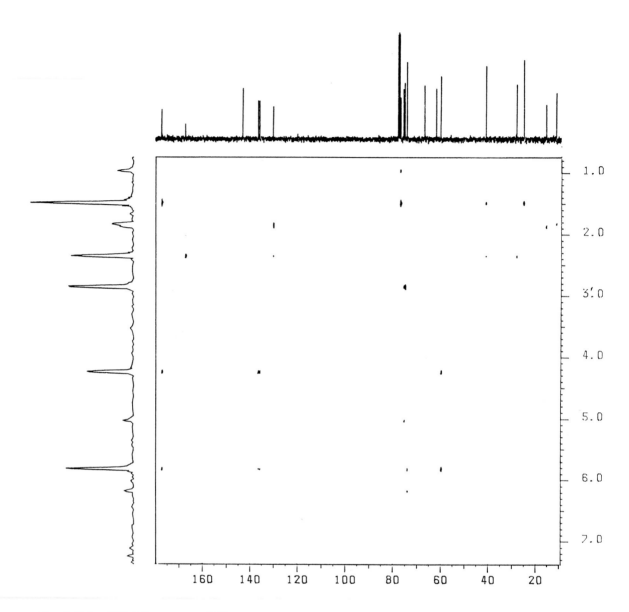

Fig. 3.18.5. COLOC spectrum of **26**.

Exercise 19

During a prostaglandin synthesis[1] a mixture of diastereomers was obtained, compound **29** was isolated from this mixture. The configuration of C–2 cannot be easily identified from the 1D ^{1}H and ^{13}C NMR spectra.

Interpret the spectra and determine whether the methoxy group at C–2 is in the α or β position. What is the preferred conformation for the tetrahydrofurane ring? Is there any information about the conformational behavior of the side chain at C–5?

Note: The substance contained a trace of an impurity.

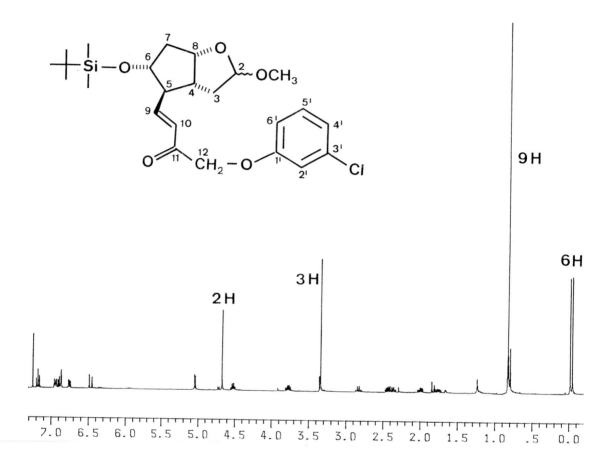

Fig. 3.19.1. ^{1}H NMR spectrum of **29**.

[1] Achmatowicz B, Baranowska E, Daniewski AR, Pankowski J, Wicha J (1985) *Tetrahedron Lett* 5597; Achmatowicz B, Marczak S, Wicha J (1987) *J Chem Soc Chem Commun* 1226.

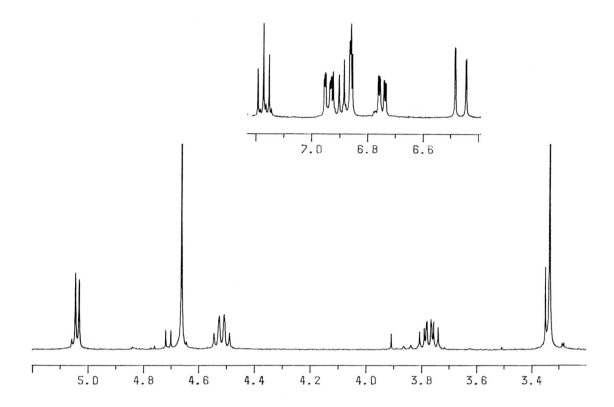

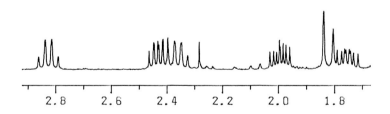

Fig. 3.19.2. Expanded sections of the ^{1}H NMR spectrum of **29**.

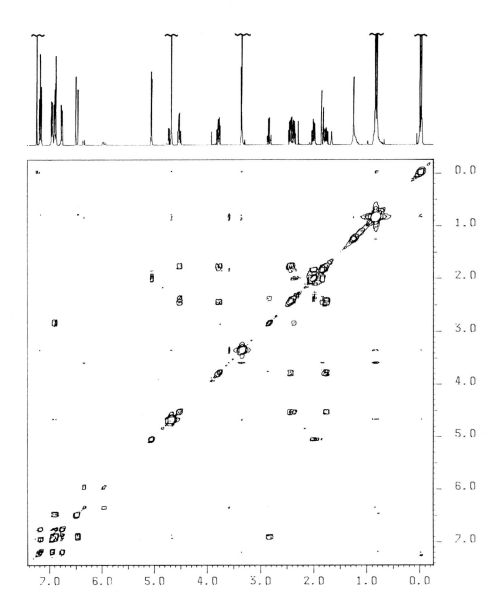

Fig. 3.19.3. H,H COSY spectrum of **29**.

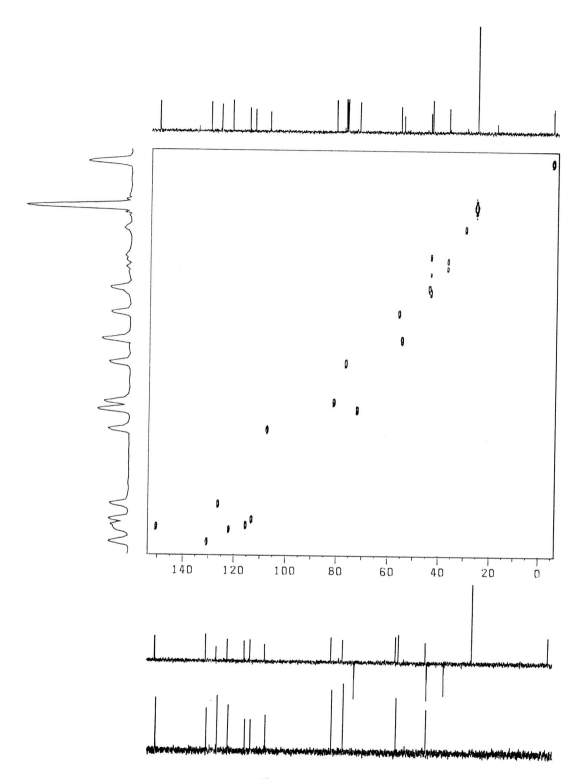

Fig. 3.19.4. H,C COSY and DEPT spectra of **29**; the ^{13}C NMR spectrum contains three additional signals at δ = 194.3 (C), 158.6 (C), and 135.0 (C).

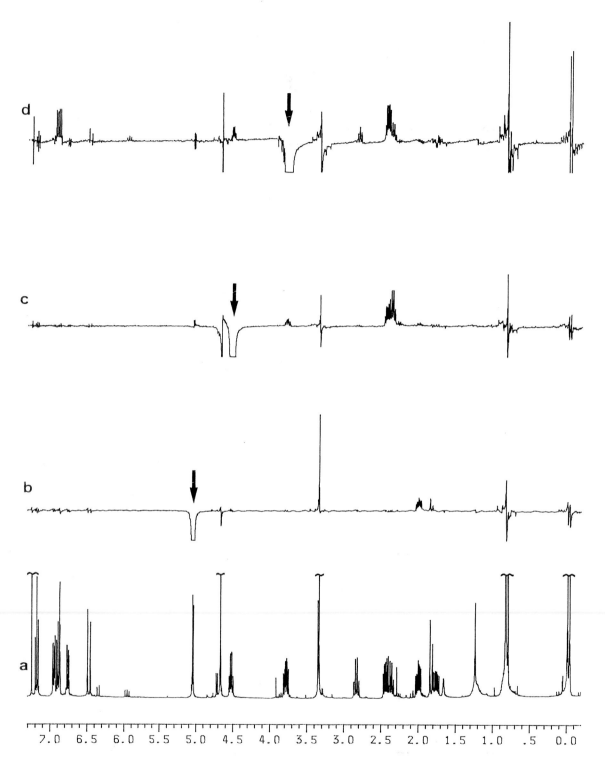

Fig. 3.19.5. NOE difference spectra of **29**: (a) Reference spectrum. Irradiarion positions at (b) $\delta = 5.03$, (c) $\delta = 4.52$, (d) $\delta = 3.77$, (e) $\delta = 2.83$, (f) $\delta = 1.98$, (g) $\delta = 1.82$.

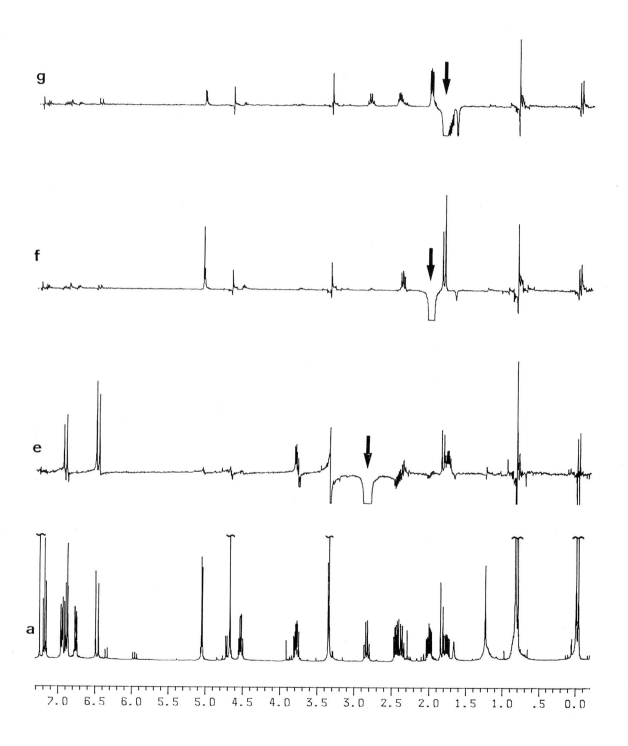

Exercise 20

A compound with the molecular formula $C_{10}H_{14}O$ has an IR band at 1740 cm^{-1}. What is its structure?

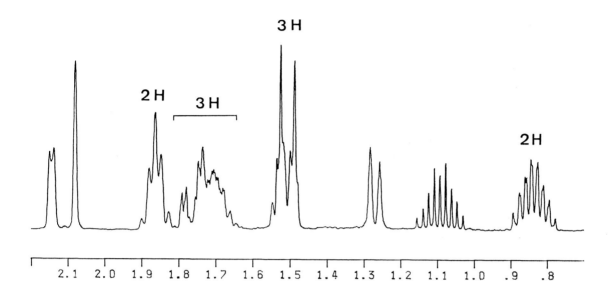

Fig. 3.20.1. ^{1}H NMR spectrum of **30**.

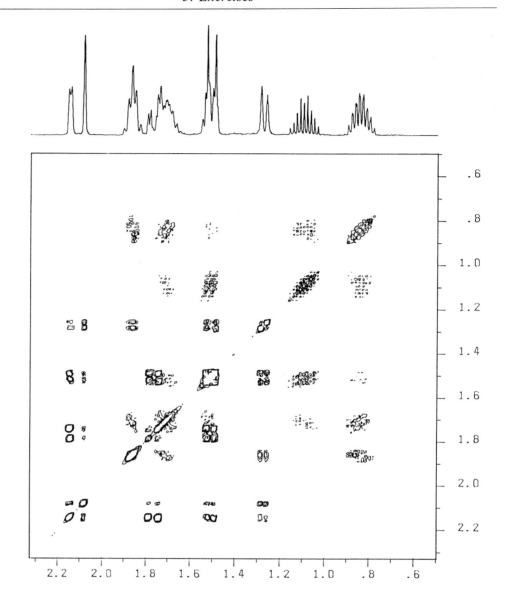

Fig. 3.20.2. H,H COSY spectrum of **30**.

Fig. 3.20.3. H,C COSY and DEPT spectra of **30**; the ^{13}C NMR spectrum contains an additional signal at $\delta = 216.8$ (C).

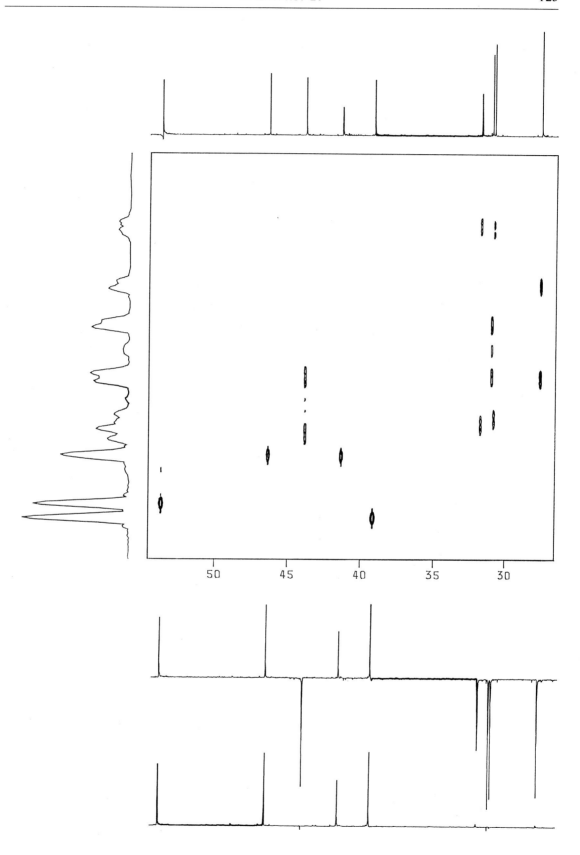

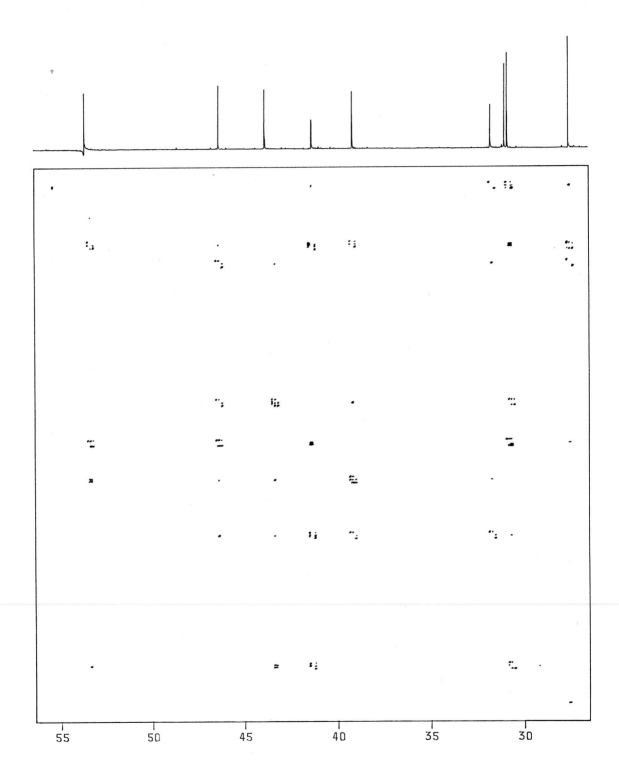

Fig. 3.20.4. 2D INADEQUATE spectrum of **30**.

Exercise 21

The diamagnetic binuclear ruthenium complex $[L_2Ru_2^{III}(\mu-O)(\mu-CH_3CO_2)_2](PF_6)_2 \cdot 0.5\,H_2O$ (31)[1] contains one oxo and two acetate bridges, as well as two 1,4,7-trimethyl-1,4,7-triazacyclononane ligands (L):

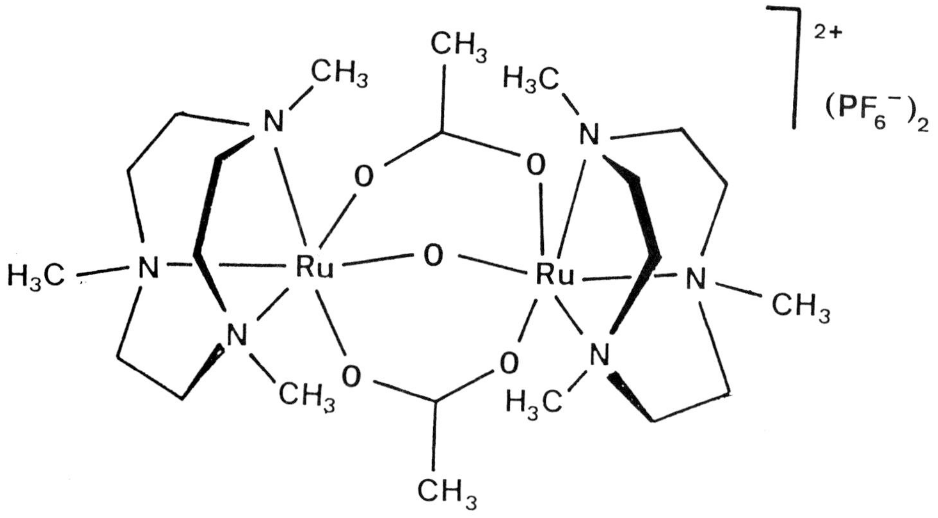

Two signals in the ^{1}H NMR spectrum resonate at unexpected frequencies. Which are the corresponding protons and what might be the reason for their unusual chemical shifts?

---

[1]The given structure was proved by X-ray crystallography: Neubold P, Wieghardt K, Nuber B, Weiss J (1988) *Angew Chem* **100**: 990; *Angew Chem Int Ed Engl* **27**: 933.

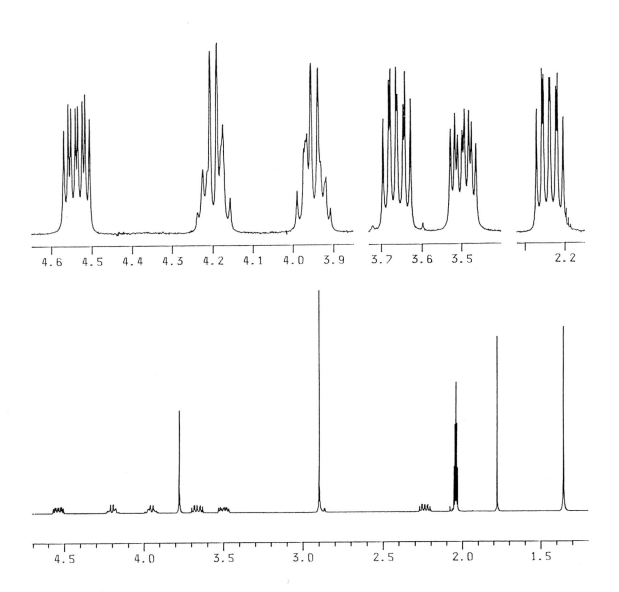

Fig. 3.21.1. ^{1}H NMR spectrum of **31**, in acetone–d$_6$.

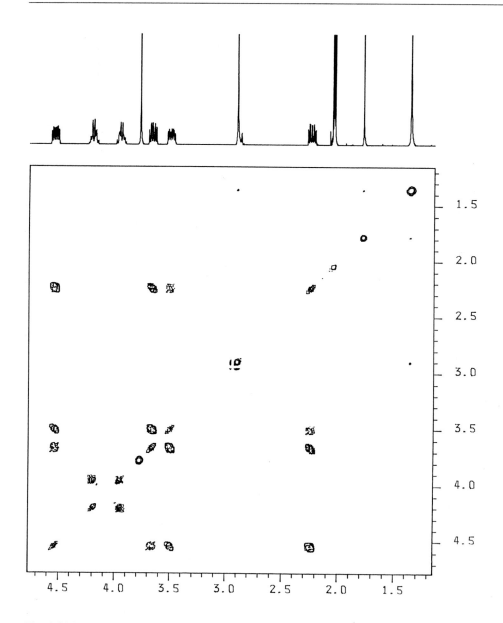

Fig. 3.21.2. H,H COSY spectrum of **31**, in acetone–d$_6$.

Fig. 3.21.3. H,C COSY and DEPT spectra of **31**, in acetone–d$_6$; the ^{13}C NMR spectrum contains an additional signal at $\delta = 193.0$ (C).

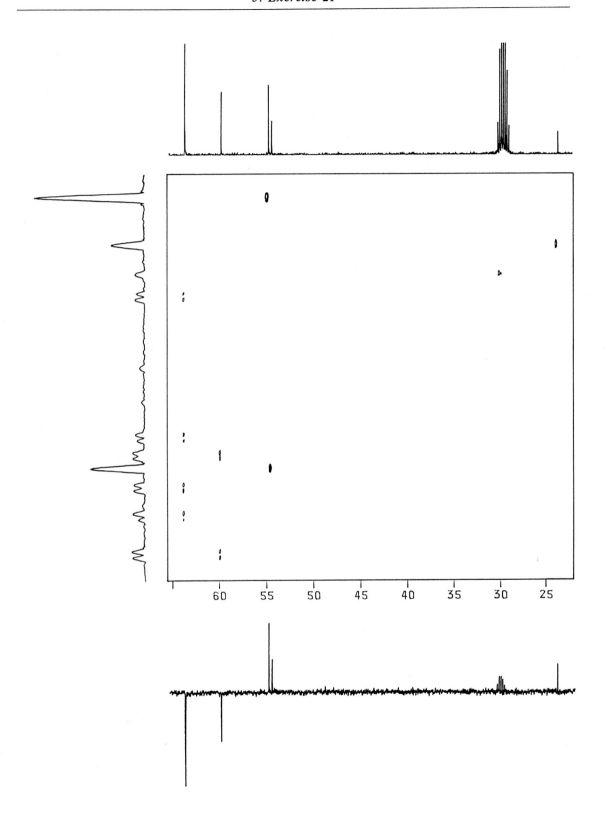

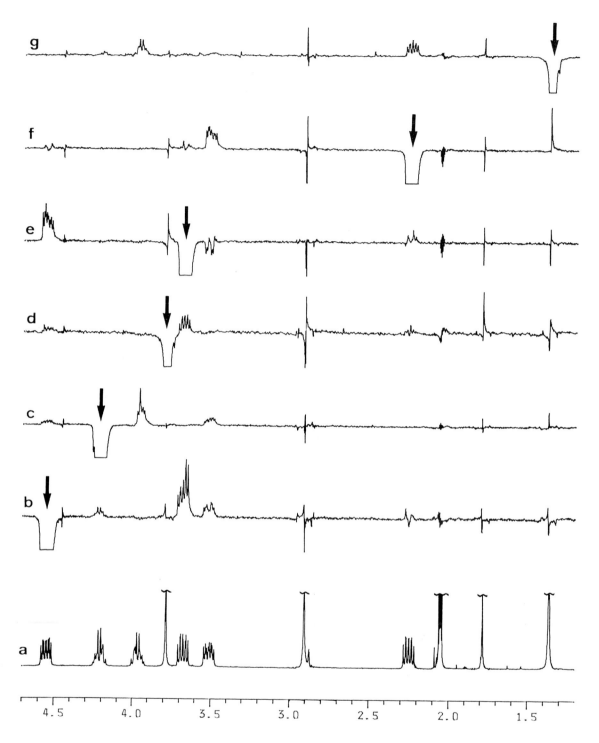

Fig. 3.21.4. NOE difference spectra of **31**: (a) Reference spectrum. Irradiation at (b) $\delta = 4.53$, (c) $\delta = 4.20$, (d) $\delta = 3.77$, (e) $\delta = 3.66$, (f) $\delta = 2.24$, (g) $\delta = 1.35$.

Exercise 22

An alkaloid **32** has the molecular formula $C_{20}H_{19}NO_5$. Its IR spectrum shows a band ranging from 3030 to 2940 cm^{-1}, even at very low concentrations. What is the structure of the alkaloid, its relative configuration, and preferred conformation?

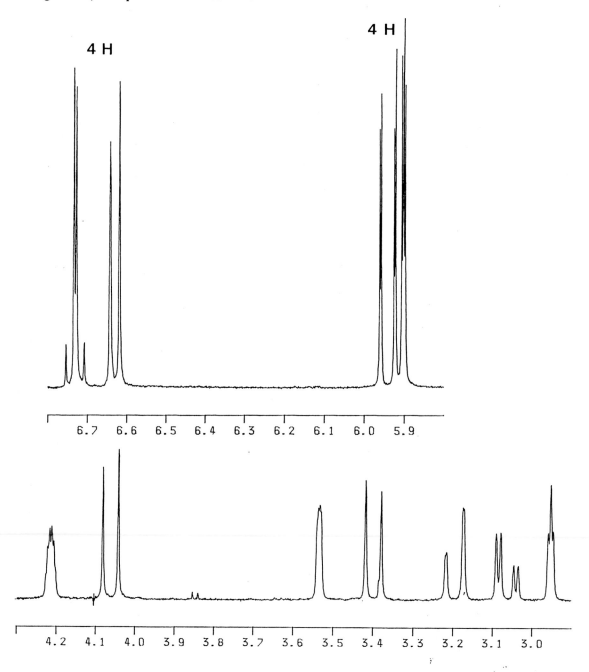

Fig. 3.22.1. ^{1}H NMR spectrum of **32**; the spectrum contains an additional broad signal at $\delta \approx 7.7$ (^{1}H) and a singlet (3H) at $\delta = 2.25$ (see also **Fig. 3.22.2a**).

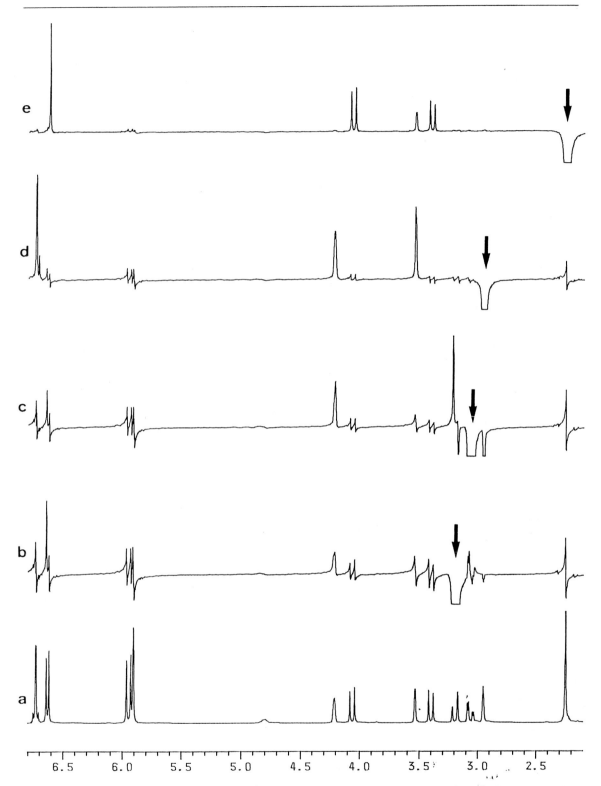

Fig. 3.22.2. NOE difference spectra of **32**: (a) Reference spectrum. Irradiation at (b) $\delta = 3.18$, (c) $\delta = 3.07$, (d) $\delta = 2.96$, (e) $\delta = 2.25$.

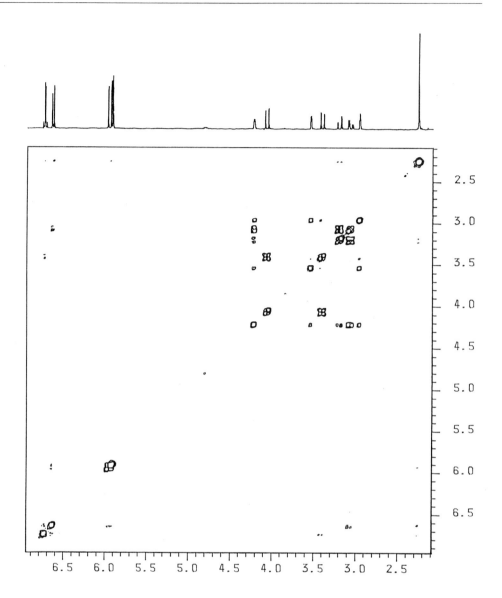

Fig. 3.22.3. H,H COSY spectrum of **32**.

Fig. 3.22.4. H,C COSY and DEPT spectra of **32**; the ^{13}C NMR spectrum contains seven additional signals at $\delta =$ 147.9, 145.4, 145.0, 142.8, 131.1, 128.6, and 125.4 (all C); see also the ^{13}C trace (horizontal) in the COLOC spectrum. The artifacts at the flanks of the H,C COSY cross peaks in the F1 dimension (^{1}H) resulted from a Gauss filtering of the FIDs in this direction in order to increase the sensitivity of the experiment.

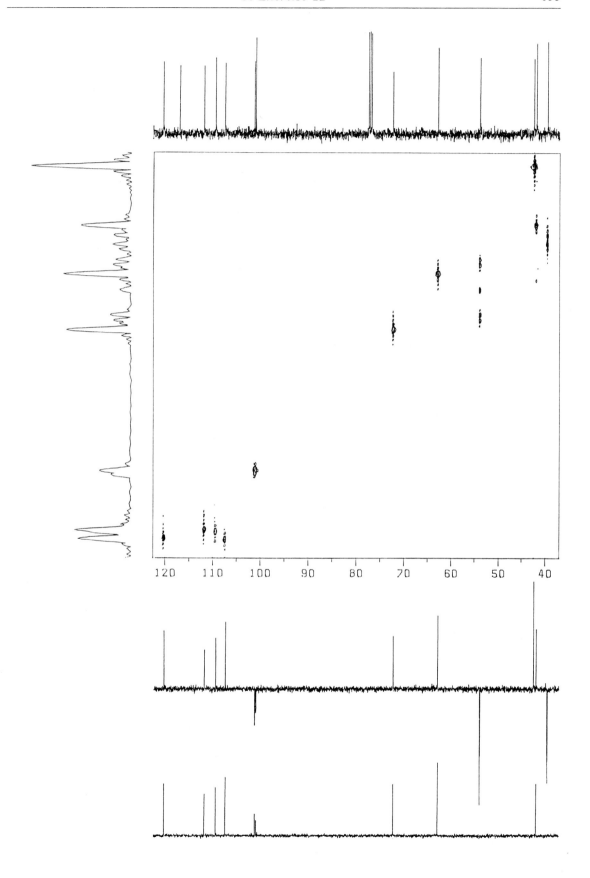

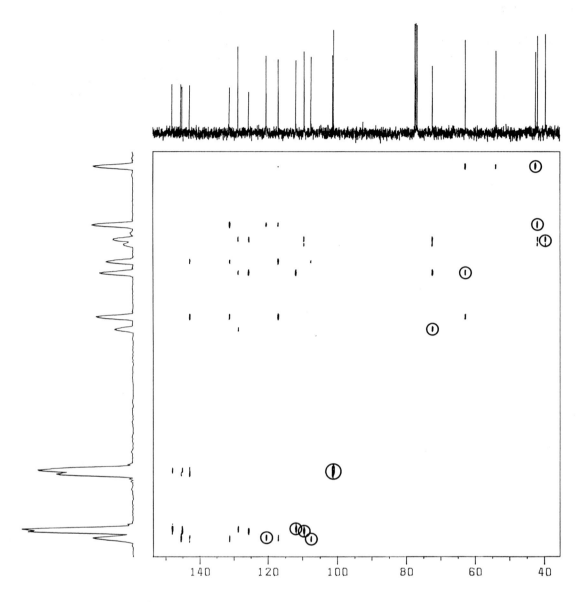

Fig. 3.22.5. COLOC spectrum of **32**; the cross peaks indicating ^{1}H,^{13}C coupling via one bond ($^1J_{CH}$, cf. H,C COSY spectrum) were circled after plotting.

Exercise 23

A natural product **33**, isolated from the plant *Melia azerarach L.*, growing in Sudan has a molecular formula $C_{28}H_{34}O_7$, established by high-resolution mass spectrometry. Moreover, its IR and UV spectra indicate the presence of an α,β-unsaturated ketone functionality and a furane ring. What is the structure of compound **33**?

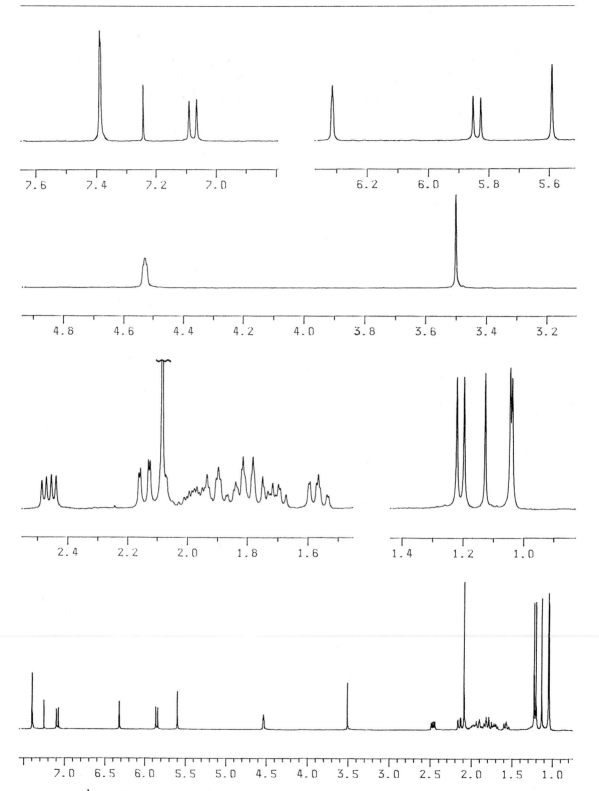

Fig. 3.23.1. ^{1}H NMR spectrum of **33**.

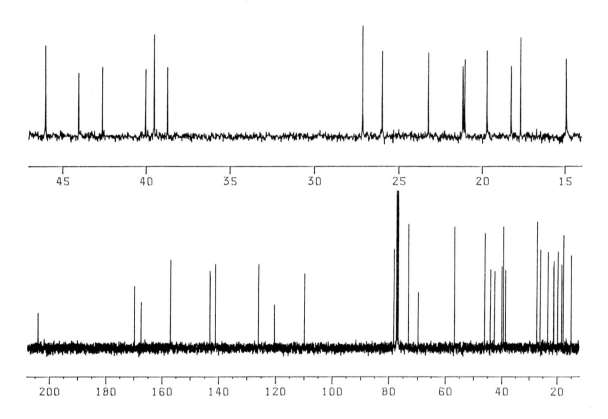

Fig. 3.23.2. ^{13}C NMR spectrum of **33**.

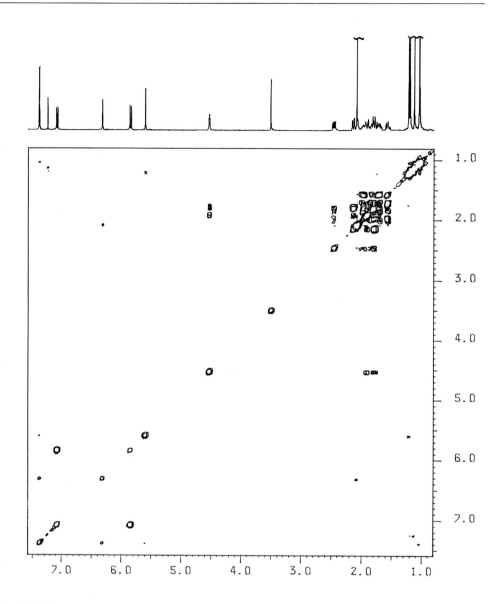

Fig. 3.23.3. H,H COSY spectrum of **33**.

Fig. 3.23.4. H,C COSY and DEPT spectra of **33**; the ^{13}C NMR spectrum contains three additional signals at δ = 204.0, 169.9, and 167.4 (all C).

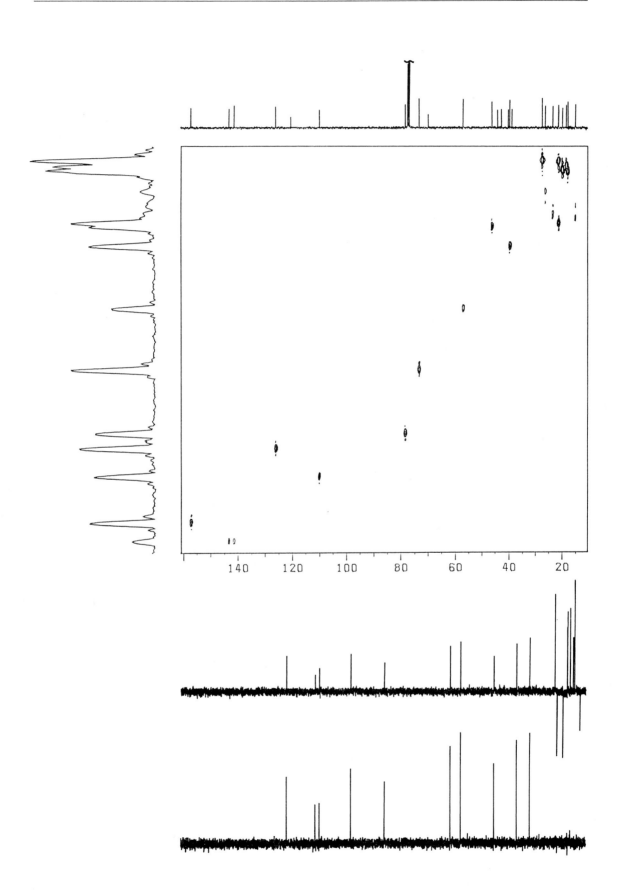

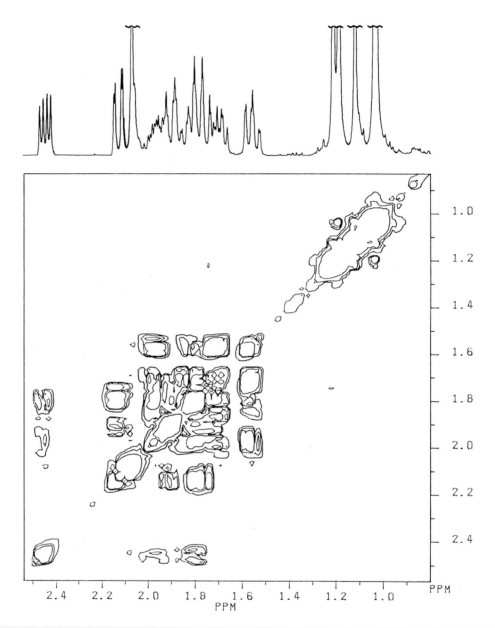

Fig. 3.23.5. Expanded section of the H,H COSY spectrum of **33**.

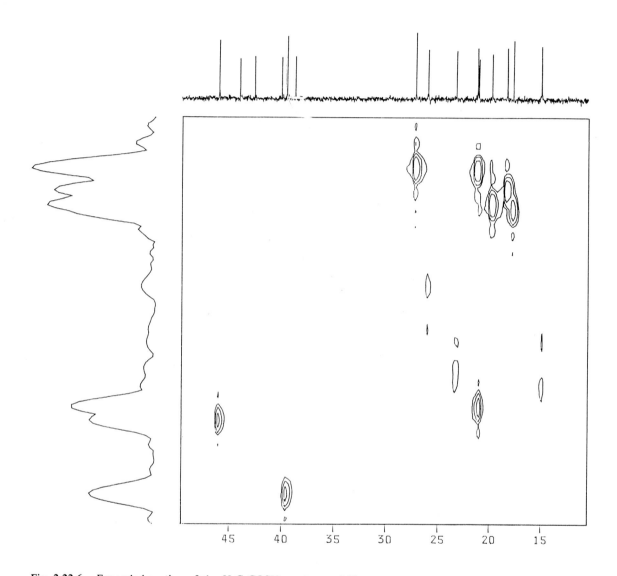

Fig. 3.23.6. Expanded section of the H,C COSY spectrum of **33**.

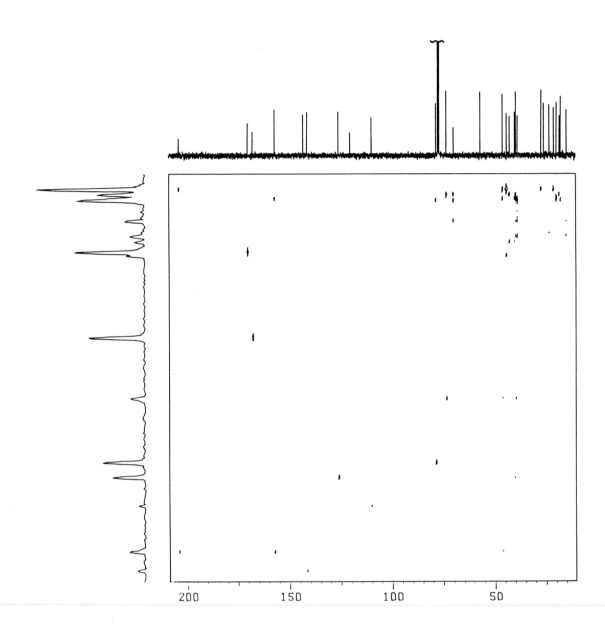

Fig. 3.23.7. COLOC spectrum of **33**.

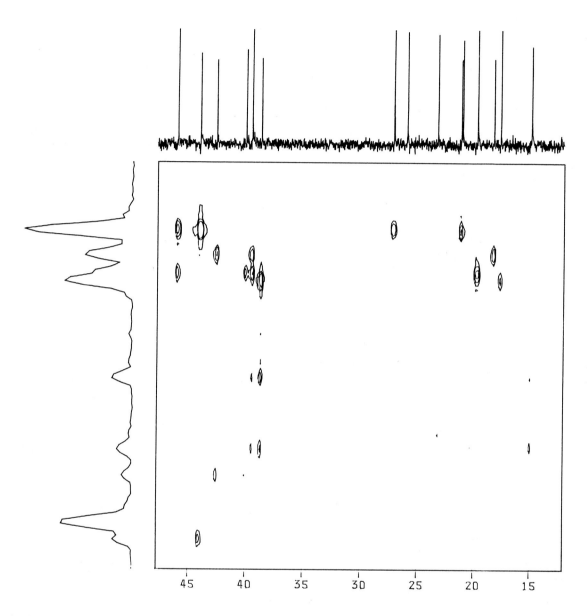

Fig. 3.23.8. Expanded section of the COLOC spectrum of **33**.

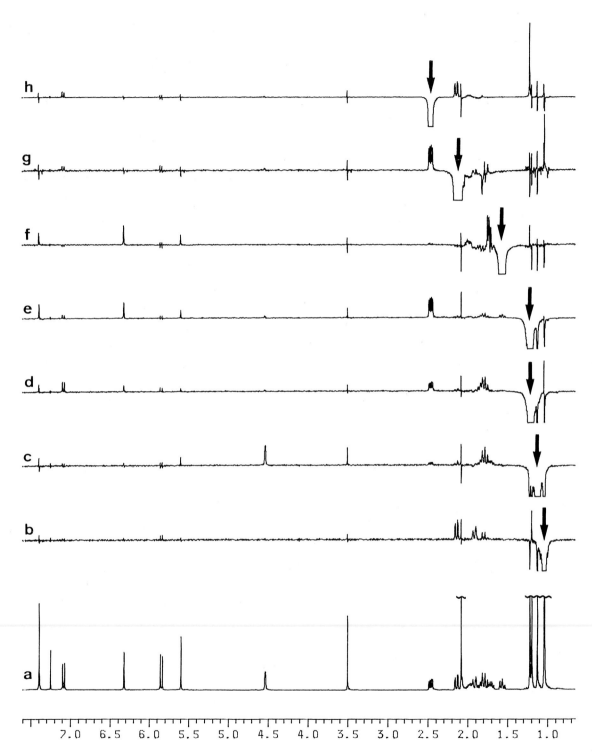

Fig. 3.23.9. NOE difference spectra of **33**: (a) Reference spectrum. Irradiation at (b) $\delta = 1.03$, (c) $\delta = 1.12$, (d) $\delta =$ 1.19, (e) $\delta = 1.22$, (f) $\delta = 1.56$, (g) $\delta = 2.12$, (h) $\delta = 2.46$, (i) $\delta = 3.50$, (k) $\delta = 4.52$, (l) $\delta = 5.59$, (m) $\delta = 5.84$, (n) $\delta =$ 6.31, (o) $\delta = 7.07$, (p) $\delta = 7.39$.

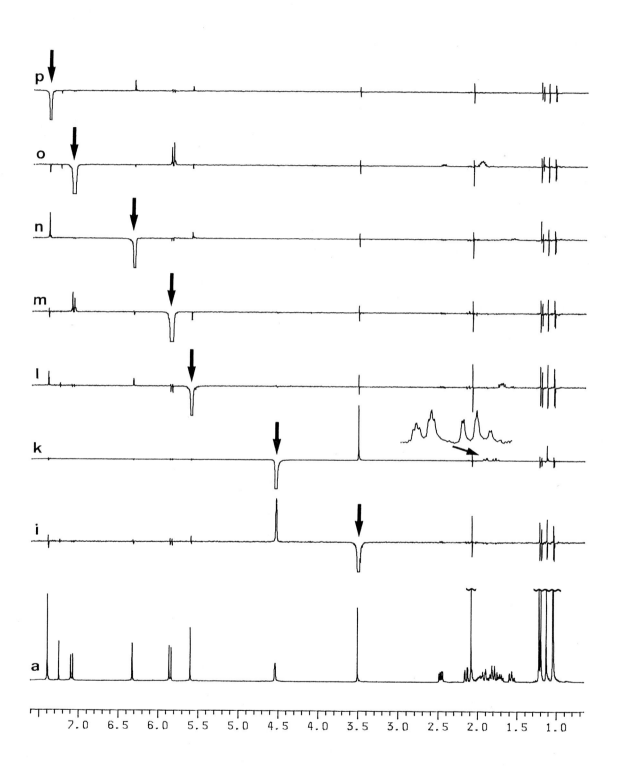

4. Strategies

The NMR spectra, especially the 2D spectra discussed in Chapter 3, contain a wealth of information. It is important, therefore, to be certain of the relative value of each single piece of information. There are "soft" clues or hints that can only be evaluated through experience and a knowledge of analogous findings, that is, on the basis of empirical evidence. A typical example is the evaluation of the chemical shift (δ) of an atom by taking into account the nature of its neighboring atoms. Such hints should, whenever possible, be verified by experimental results. Information from characteristic δ ranges are more reliable. For instance, is clear that a ^{13}C signal with $\delta = 210$ corresponds to a carbonyl carbon and that this carbon cannot be a member of a carboxyl function. Many ^{1}H,^{1}H and ^{13}C,^{13}C coupling constants belong to the same category of reliable hints.

In contrast, many modern NMR techniques afford unequivocal "hard" proofs, which should always form the basis of any interpretation. For example, cross peaks in COSY spectra prove the existence of scalar coupling and thus render a statement concerning the connectivities of atoms within the molecule. The appearance or absence of nuclear Overhauser effects (NOE difference signals) can be considered conclusive with regard to the spatial relations of the respective nuclei, within the given limits of reliability for this method.

It is not easy to obtain an overall view with respect to all the many bits of information that must be evaluated for relevance. Therefore, it is essential to prepare a strategy at the very beginning of a spectral interpretation. Such a strategy, of course, depends strongly on information already available from other sources.

If the problem is mainly a question of signal assignment, that is, if the molecular structure is already known (e.g., Exercises 3, 16, and 21), one should start with a signal that can be easily assigned because of its chemical shift or its coupling pattern. Often this predication is based only on a fairly reliable clue that has to be verified during the course of subsequent interpretation. If, at the end, the bits of information cohere and the series of arguments prove conclusive, the initial assumption was correct. Fortunately, there are often several independent experimental clues and/or proofs for a particular piece of information, a situation that improves as the complexity of the molecule increases.

In some exercises only the constitution of the compound (that is, the atomic connectivity without any stereochemistry) or fragments of the molecule are known (e.g., Exercise 6, 9, 10, 14, 18, and 19). In such cases it is advisable first to assign the signals if and to the extent possible and second to determine the stereochemical details, such as configuration and/or conformation.

The most difficult task is an exercise without any structural information, except for some general statement, a molecular formula, for example. Here, a strategy must be devised for establishing molecular fragments, and the search for experimental facts by a combined and simultaneous evaluation of all 1D and 2D spectra should be the foremost concern from the outset. Moreover, it is important to find a suitable "entry point," that is, a signal corresponding to an atom or group of atoms for which the assignment is straightforward. Often proton and carbon signals of methyl groups or C_1 fragments containing an oxygen functionality are most suitable. In saccharide spectra the best entry points are signals of the anomeric carbon/hydrogen or the terminal $-CH_2OR$ group. Once a good starting point is found, the spectral evaluation should first be restricted to determining the constitution of the unknown compound. Stereochemical details should be postponed to a later stage because, to a considerable degree, such information will be derived from NOE evidence. Nuclear Overhauser effects, however, may be misleading in establishing the molecular skeleton, since they respond to the spatial proximity of atoms and not to their relationships with their neighbor in terms of intervening bonds.

An important aid in finding structural fragments is the molecular formula, since it provides information about the number of atoms still to be identified. In addition, it is very useful for determining the number of unsaturated equivalents (double bonds or rings), which tell us how many rings have to be formed by combining the previously identified fragments with the constitution formula. Such an approach is very much like assembling a jigsaw puzzle – units are put together piece by piece until they finally combine to form the total picture.

Next, for each exercise we offer some advice about strategy to help the inexperienced reader solve the problem. By no means, however, does this mean that the suggested strategy is the only possibility. It is intended to orient and encourage the reader in his or her own attempt to arrive at a step-by-step solution. It is suggested that the reader considers this preliminary material in each exercise carefully before going on to the explicit procedures described in Chapter 5.

Exercise 1

It is already possible to identify some functional groups in the 1D spectra by inspecting the ^1H and ^{13}C chemical shifts. The two COSY spectra permit the combination of these groups into molecular fragments that eventually can be combined into the constitution formula. The signal can be assigned by evaluating one particular cross peak in the H,H COSY spectrum indicating the existence of a long-range coupling.

Exercise 2

First, the number of double-bond equivalents should be determined, since comparing its number and the number of detectable double bonds affords information as to whether or not **2** is cyclic. The position of the double bonds can be found by establishing the $^1H,^1H$ connectivity using the H,H COSY spectrum.

To determine the stereochemistry of **2**, a selective ^1H decoupling experiment is appropriate, since coupling constants may be extracted from the splitting of the olefinic proton signals. An NOE difference experiment may also be useful for recognizing the spatial proximity of the olefinic protons and/or those in aliphatic CH_n fragments directly attached to the double bond.

In both experiments a hydrogen atom neighboring a double bond should be irradiated.

Exercise 3

It is advisable to start with the ^{1}H signal assignment of protons in the CH fragments. Thereafter, the methylene protons can be identified. Their stereochemical position can be determined by evaluating ^{1}H,^{1}H long-range couplings in W arrangements (^{4}J$_{HH}$), since only in this coplanar orientation can cross peaks of significant intensity be observed for couplings at this distance.

The multiplicities of the proton signals can be evaluated as well supporting previously obtained arguments.

Finally, the H,C COSY spectrum provides the ^{13}C signal assignment.

Exercise 4

The ^{1}H,^{1}H connectivity can be established as proposed in Exercise 3. The multiplicities of the proton signals from the CHBr fragments provide information about the number of hydrogen atoms in the W position.

For a check of the ^{13}C signal assignments, increment rules[1] can be used to calculate the ^{13}C chemical shifts of **5** through **7**. One of them should provide a good fit with the experimental values.

Exercise 5

Both 1D spectra clearly reveal that the substance is a mixture containing a major and a minor constituent.

It is first necessary to sort out the signals belonging to each of the components by using the COSY spectra. Next, it should be determined whether the two compounds are structurally different or if they are stereoisomers. The constitution can be determined by establishing atom connectivities through inspection of the COSY spectra. The stereochemistry can be investigated by means of an NOE difference experiment, and the ^{13}C chemical shifts may confirm these findings.

Exercise 6

There are three chirality centers in the molecule, namely, C–6, C–6a, and C–7, so that four different pairs of diastereomers are conceivable. Moreover, the seven-membered ring may undergo a conformational ring inversion. The key signals for determining the stereochemistry of **12** are those of the protons attached to the chirality centers (H–6, H–6a, and H–7) absorbing between $\delta = 5.1$ and 3.6. After the signals are assigned, their splittings provide coupling information that permits conclusions regarding the relative arrangement of the hydrogens and the torsional angles between them.

---

[1] The increment rules have been derived from experimental data for brominated adamantanes and adamantanones: Duddeck H (1975) *Org Magn Reson* **7**: 151.

Exercise 7

The mass spectrum indicates the molecular weight. First, the number of double bond equivalents (double bonds or rings) have to be determined. Since the molecular weight is low, rings should be small in size. Cyclopropane rings can be identified by the relatively large one-bond carbon-hydrogen coupling $^1J_{HH}$). The heteronuclear NOE experiment (Fig. 3.7.2c) provides additional help in making signal assignments.

Exercise 8

Certain structural fragments can be derived from the chemical shifts, coupling constants, and cross peaks in the COSY spectra. Arguments about local symmetry should be followed when fragments are being combined into a structural formula. The NOE difference experiment is helpful in making further signal assignments.

Exercise 9

This exercise is confined to signal assignments for ring D, that is, the five-membered ring in the steroidal skeleton, and for substituents attached to it. This proton spin system is isolated from the rest of the molecule by the two quarternary carbons C–13 and C–14, so a complete ^1H and ^{13}C signal assignment for atoms in or at the other six-membered rings A to C is not essential.

With the help of the two COSY and the DEPT spectra, the signals of H–15 to H–20, H–23, and H–24 can be identified. The NOE experiments allow an unambiguous determination of the stereochemistry of this part of the molecule. A discussion of the signal splittings provides further arguments.

Exercise 10

The enone part is separated from the rest of the molecule by two quarternary carbons (C–4 and C–10). So, the question of whether the compound has structure **17** or **18** cannot be answered by an H,H COSY spectrum since the $^1H,^1H$ connectivity is interrupted. Therefore, a COLOC spectrum has been recorded because it proves $^1H,^{13}C$ couplings via more than one bond.

Possibly an NOE difference spectrum might be helpful if spatial proximities between the olefinic protons and those of methyl groups 14 and 15 can be identified. This, however, requires a full and unambiguous signal assignment, which can be achieved through the two COSY spectra. As expected, the most informative source for determining the stereochemistry of **16** is again the NOE difference experiment.

Exercise 11

First, it is necessary to determine the ^{1}H,^{1}H and ^{1}H,^{13}C connectivities by evaluating the COSY spectra. This will tell what kind of substituents are present. The NOE difference spectra provide information about the stereochemistry of **19**.

Exercise 12

The compound is highly unsaturated and seems to contain a carboxylic acid and a phenol function. Establishment of the ^{1}H,^{1}H connectivity is hampered by the small number of hydrogen atoms and an overlap of several signals at $\delta = 2.2$ to 2.0. This molecule is an excellent candidate for a COLOC experiment. Finally, an NOE difference experiment confirms stereochemical and methyl signal assignments and gives surprising results concerning segmental motions within the molecule.

Exercise 13

By determining the ^{1}H,^{1}H connectivity in the sugar part and evaluating the ^{1}H,^{1}H coupling constants, we can derive the structure of the saccharide. The question regarding the stereochemistry of the anomeric carbon can be answered by inspecting the H–1 signal. The signal splittings of the aromatic hydrogens allow us to determine the structure of the aglycone; the latter can be easily checked by calculating ^{1}H and ^{13}C chemical shifts using increment rules.

Exercise 14

Our first consideration is the symmetry of the molecule, that is, which stereoisomers are consistent with the spectra. We can establish the ^{1}H,^{1}H connectivity using the two COSY plots. The NOE difference spectrum provides the stereochemistry of **22**. In addition, it will be helpful to determine the reason why one proton signal has an unusually small chemical shift ($\delta = 0.75$).

Exercise 15

Besides the two already mentioned aromatic rings with their substituents, the molecule contains three additional carbons, four hydrogens, and two oxygens. Thus, the first task is to combine these structural fragments into a constitution formula. The number of double-bond equivalents should be considered.

In order to identify the positions of the substituents (acetate and methoxy), the spin systems of the two independent aromatic units must be assigned. This assignment can be made on the basis of the NOE difference experiments.

Exercise 16

The greatest problem in assigning the ^{13}C signal of **24** is the differentiating among the three nearly isochronous olefinic methine carbons with $\delta = 132.4$, 132.2, and 132.1. Such a differentiation cannot be achieved on the basis of the COSY spectra because of their limited resolution; only a 2D INADE-QUATE spectrum, from which the ^{13}C,^{13}C connectivity can be obtained directly, can help.

A good entry point for establishing the connectivity table is the identification of the two methylene groups and their neighbors. The assignment of the three above-mentioned olefinic carbons requires evaluation of the expanded section of the 2D INADEQUATE spectrum.

The H,C COSY plot allows a ^{1}H signal assignment, which should be verified by inspection of the H,H COSY spectrum.

Exercise 17

The number of ^{13}C signals in the region of $\delta = 50$ to 105 tells how many monosaccharide units are present. In the H,H COSY spectrum the ^{1}H signals can be attributed to the respective units, and the proton signal splittings provide information about the nature of the sugars. The H,C COSY makes possible the ^{13}C signal assignments within the monosaccharide moieties. The selective INEPT experiment (Fig. 3.17.2) provides an unequivocal proof for a connectivity between the anomeric hydrogen of one monosaccharide unit and the glycosidated carbon of the other.

Exercise 18

The first step is the identification of the ^{1}H and ^{13}C signals and the determination of the ^{1}H,^{1}H connectivity within the two fragments **27** and **28** by evaluating the two COSY plots. The COLOC spectrum is helpful in finding the connection between the two molecular parts because it provides information about ^{1}H,^{13}C long-range couplings across the connection sites, especially the carbonyl carbons.

Thereafter ^{1}H,^{1}H coupling constants and NOE signal enhancements can be evaluated to obtain information about the spatial orientation of the protons in the macrocyclus. It is possible to find the relative configuration and to see that there is a preferred conformation.

Exercise 19

The evaluation of ^{13}C chemical shifts and ^{1}H,^{1}H coupling constants in five-membered rings is often difficult because of their molecular flexibility. To answer the question regarding the stereochemistry at C–2, it is essential to determine the orientation of H–2 relative to the two H–3 atoms by using NOE difference spectra. First, of course, an unambiguous stereochemical assignment for these hydrogens is needed, and it can be accomplished through the H,H COSY and some NOE difference spectra.

Exercise 20

Apparently, there are no olefinic or aromatic carbons in the molecule. In agreement with the molecular formula, the ^{13}C NMR spectrum displays ten signals; so the compound must be an unsymmetrical tricyclic ketone. The IR band permits conclusions about the size of the ring in which the carbonyl group is situated.

We could begin by inspecting the two COSY plots to establish the connectivities of the atoms. Unfortunately, this approach would not prove successful, because of the severe overlap of important protons. Only a 2D INADEQUATE spectrum can provide the required information.

Exercise 21

This exercise requires an unequivocal proton signal assignment in the triazacyclononane ligand. Owing to symmetry the proton spin systems are different. We can use one of them or the N-methyl signals as entry points. NOEs help us to find ^{1}H,^{1}H connectivities across the nitrogen atoms.

For an explanation of the unusual proton chemical shifts, it is necessary to consider the positions of the involved nuclei with respect to the central oxygen atom.

Exercise 22

The highly unsaturated compound **32** contains relatively few olefinic/aromatic hydrogen atoms, but many sp^2 carbon atoms, eight of which are quarternary. Thus, it is reasonable to assume that **32** contains several polysubstituted aromatic rings. In addition, seven aliphatic hydrogens can be identified, and their connectivity easily determined. From the ^{1}H,^{1}H coupling constants their relative spatial orientation can be estimated. The NOE difference spectra are helpful in interconnecting the aliphatic and unsaturated fragments. The ^{1}H and ^{13}C signals in the oxygen functionalities display unusual chemical shifts. There are several possible stereoisomers with different configurations and conformations. We must determine which of them is consistent with the spectral data before we can arrive at the final structure.

Exercise 23

This last exercise is admittedly very difficult, since it deals with a complex and largely aliphatic molecule containing 28 carbons. Nine of them are quarternary, so the skeleton falls into a number of protonated fragments separated from each other by these quarternary carbons and/or oxygen. First, such fragments have to be identified. Then it is necessary to find a suitable entry point by which they can be joined together. Finally, NOEs are used to help in determining the sophisticated stereochemisty of the molecule.

In our opinion, with the limited arsenal of methods introduced in this book, the complexity of this example approaches the application limits for chemists who are not NMR specialists. We believe that at this time structural elucidations of greater complexity will remain the province of experts and/or, in the foreseeable future, extend to computer-aided interpretation.

5. Solutions

In this chapter we suggest explicit procedures for solving the problems in the 23 exercises. It is again emphasized that the proposed solutions are by no means the only ones possible. Others may be better and more elegant.

It is not our intention to provide a comprehensive evaluation of all the possible information inherent in the spectra. Thus, the reader who decides to go along with the proposed solutions should try to obtain additional supporting evidence.

In many cases it is highly advisable to use molecular models, especially when NOEs are discussed in terms of through-space interatomic distances. Here human imagination – particularly when coupled with lack of experience – quickly reaches its limits.

Sometimes it is not possible to assign all signals unequivocally and without relying on "soft" clues. Often, however, such a limitation is acceptable, that is, for a satisfactory answer to the problem a complete and absolutely safe signal assignment is not necessary. In such exercises, therefore, we have omitted experiments requiring the kind of subtle spectral evaluation that has to be paid for with an unwarranted expense of spectrometer time and personal effort, an omission typical of everyday laboratory routine.

The more complex the molecule and the higher the number of spins, the sooner we reach a point at which spectral evaluation becomes difficult. Such cases require a well-defined structure in the sequence of arguments, as discussed in Chapter 4. There then arises the question of how to document the results of our evaluation. In this chapter we attempt to establish a logical procedure, which involves first recognizing functional groups through their partial structures, the way they combine to form larger units, and their constitution formulas, and then, subsequently determining their stereochemistry. The documentation method outlined in this chapter is appropriate for a workbook and may also be suitable for master's and doctoral theses. For published reports and journal articles, however, the procedure is too detailed and voluminous, especially if several compounds have to be described.

Of course, we cannot provide a general outline for a method of documentation that is both concise and comprehensive, and the literature, unfortunately, does not offer much help[1]. The compilation of spectral evidence in tabular or graphic representation is advisable in preparing manuscripts. This type of documentation is presented in Exercises 22 and 23, in addition to the usual descriptions found for all the other exercises. For instance, cross peaks in COSY spectra can be given in matrix form (cf. Tables 5.22.1, 5.22.2, and Figs. 5.23.3, and 5.23.4); structural fragments can be arranged in blocks, as shown in Figs. 5.22.4 and 5.23.11; NOE evidence can be noted in a matrix, as in Fig. 5.23.8, or in structural graphs, in which arrows indicate the proton-proton interactions (cf. Fig. 5.22.7 and 5.23.8).

---

[1] A particularly successful example is Mascagni P, Gibbons WA, Asres K, Philipson JD, Niccolai N (1987) *Tetrahedron* **43**: 149.

Under the heading "signal assignments" a compilation of all available chemical shifts can be found at the end of most sections. In other instances these data are arranged in a table. Moreover, we often cite references that do not necessarily refer to NMR literature only: they may give more information about the substances themselves or the class of compounds to which the substance under investigation belongs.

Exercise 1

In the olefinic range of the ^{1}H NMR spectrum, three signals can be identified at $\delta = 6.56$ ($J = 9.7$ and 15.9 Hz), 5.99 ($J = 15.9$ Hz), and 5.44 (broad singlet). According to the H,H COSY spectrum the first two are coupling partners, but the third is isolated. It can therefore be concluded that the three hydrogen atoms belong to different double bonds having the constitution formulas $-CH=CH-$ and $>C=CH-$.

The 15.9-Hz coupling between the two protons proves that the first double bond is trans-configurated. A connectivity between the proton with $\delta = 6.56$ and the aliphatic proton with $\delta = 2.23$ (doublet) is revealed by a 9.7-Hz coupling. The latter signal belongs to a CH fragment ($\delta_C = 54.2$). The second proton in this double bond is weakly coupled to the hydrogens of the methyl group with $\delta = 2.20$, a ^{1}H chemical shift that is typical for an acetyl group ($\delta_C = 26.8$)[1]. The ^{13}C chemical shift of the carbonyl signal ($\delta = 198.4$) indicates the existence of an unsaturated ketone so that it can be assumed that the molecule contains a side chain of the following structure:

No further couplings for the CH proton can be detected; this group is apparently situated between two quaternary carbon atoms. The molecular formula of the residual molecule is C_8H_{14}; it is C_9H_{15} if the terminal methine fragment of the side chain, which possesses two more valences is added. This formula corresponds to two double bond equivalents, and one of them is the second double bond mentioned above. Consequently, the presence of one more equivalent shows that a ring must exist in the molecule. The third olefinic proton ($\delta = 5.44$) is coupled to methyl protons with a chemical shift of $\delta = 1.51$. Since this signal has only scarcely visible splitting, the underlying coupling could be an allylic coupling in a molecular fragment:

The remaining methyl groups ($\delta_H = 0.87$ and 0.78) are isolated; apparently, they are geminal substituents attached to a quaternary carbon atom ($\delta_C = 32.4$). Only the methyl protons absorbing at higher frequencies reveal a weak coupling with the proton at $\delta = 1.41$, which, along with a proton having a δ value of 1.17, belongs to a methylene group that neighbors a second methylene group with

---

[1] The assignment of C-10 to the signal at $\delta = 26.8$ instead of the signal at $\delta = 26.7$ cannot be verified in the H,C COSY spectrum depicted here because of its reduced scale. There would be no problem, however, if the plot were expanded, for instance, to 20 × 20 cm.

two nearly isochronous protons (δ = 2.00 to 1.98). Finally, the latter protons are coupled to the olefinic proton in the trisubstituted double bond so that the ring is closed in this position. After the obtained molecular fragments are combined, compound **1** turns out to be α-Ionon:

The weak cross peak, indicating a long-range coupling between the methyl protons with δ = 0.87 and the methylene proton with δ = 1.41, allows a stereochemical assignment. Such a coupling is conceivable only if the atoms in question are in W orientation, a condition that holds only for H–5^{qa} and H–12[2]. This assignment, however, is based on a not very convincing argument, since the weak cross peak cannot be differentiated safely from an artifact. An observation of an NOE could prove a spatial neighbor relationship between H–5^{qe} and H–12 or H–5^{qa} and H–13, respectively[2]. Because of the small difference in chemical shift between the hydrogen atoms concerned, such an experiment is not easy. Moreover, it is not essential to determine the answer of the question what the structure of compound **1**.

Signal assignment for **1**: δ_H = 2.23 (H–1), 5.44 (H–3), 2.00–1.98 (two H–4), 1.41 (H–5qa)[2], 1.17 (H–5qe)[2], 6.56 (H–7), 5.99 (H–8), 2.20 (H–10), 1.51 (H–11), 0.87 (H–12), and 0.78 (H–13); δ_C = 54.2 (C–1), 131.8 (C–2), 122.6 (C–3), 22.9 (C–4), 31.1 (C–5), 32.4 (C–6), 149.0 (C–7), 132.2 (C–8), 198.4 (C–9), 26.8 (C–10), 22.7 (C–11), 27.7 (C–12), and 26.7 (C–13).

---

[2] Designation: "qa" denotes *quasi-axial*, "qe" *quasi-equatorial*.

Exercise 2

The molecule contains four double-bond equivalents, one of which belongs to a carbonyl group and whose ^{13}C chemical shift ($\delta = 164.5$) suggests that it is an α,β-unsaturated carboxylic acid derivative. The ^{13}C NMR spectrum displays four additional olefinic CH signals corresponding to two double bonds. Thus compound **2** is monocyclic.

The H,C COSY spectrum shows that the ^{1}H signal at $\delta = 5.18$ does not belong to the double bonds but rather to a $-CH-O-$ fragment, since the corresponding ^{13}C chemical shift is $\delta = 73.8$. This structural element is centrally positioned in the molecule. Starting here, one can proceed in two different directions in the H,H COSY spectrum:

a) Via an olefinic proton ($\delta = 5.50$), a second proton with $\delta = 5.60$ is reached, leading to a signal between $\delta = 2.05$ and 2.00 that corresponds to two hydrogens. We can continue in this direction via the signals at $\delta \approx$ ca. 1.33 (2H) and 1.3 to 1.2 (4H) to a terminal methyl group with $\delta = 0.84$. Thus, the entire fragment is a side chain containing one double bond.

b) At the other side of the $-CH-O-$fragment, we first have a methylene group ($\delta = 2.34$ and 2.31), which leads to a second double bond. This and the $-(C=O)-O-$fragment have to belong to the cyclic substructure. Since the carbonyl group, as mentioned above, is part of an α,β-unsaturated carboxylic acid derivative and the central methine fragment apparently bears an oxygen functionality, the constitution formula for **2** is the following:

The configuration of the double bond can be easily determined by double-resonance experiments: Irradiation of the H–9 atoms ($\delta = 2.05$ to 2.00) in a selective decoupling experiment affords a simplification of the H–8 signal, and from the residual splitting a coupling constant $^3J(7, 8) = 10.5$ Hz can be recognized. Thus, the double bond is cis or Z configurated. The NOE difference experiment gives the same result: The enhancement of the H–8 signal is, of course, pronounced, whereas the NOE signal for H–7, which is in the trans-position with respect to H–9 is extremely weak, that of the cis-oriented H–6, however, is significant. Some of the partial signals of H–8 are hardly visible or even negative. This originates from the fact that the H–8 and two H–9 nuclei are connected by a significant coupling (cf. Sect. 2.3). Compound **2**, named argentilactone [1,2], has the following structure:

The H–11 and H–12 nuclei are nearly isochronous. Therefore, it is hardly possible to establish an ^{1}H,^{1}H connectivity that can be used for the assignment of the C–11 and C–12 signals in the H,C COSY spectrum, which are 9 ppm apart. This large ^{13}C chemical shift difference, however, allows a differentiation. Since C–12 has only one β carbon (C–10) whereas C–11 has two (C–9 and C–13), the δ value of the latter is expected to be larger by about 9 to 10 ppm. Increment rules well known in ^{13}C NMR literature [3] can reproduce not only this difference but even the absolute δ values of C–11 and C–12, with satisfactory precision. With this knowledge it is now possible to recognize that the H–11 protons have chemical shifts slightly larger than those of H–12 in the H,C COSY spectrum.

Signal assignment for **2**: δ_H = 5.99 (H–3), 6.84 (H–4), 2.34 and 2.31 (two H–5), 5.18 (H–6), 5.50 (H–7), 5.60 (H–8), 2.05–2.00 (two H–9), 1.33 (two H–10), 1.25–1.20 (two H–11 and two H–12), and 0.84 (three H–13); δ_C = 164.5 (C–2), 121.4 (C–3), 144.9 (C–4), 29.8 (C–5), 73.8 (C–6), 126.3 (C–7), 135.6 (C–8), 27.7 (C–9), 28.9 (C–10), 31.3 (C–11), 22.3 (C–12), and 13.9 (C–13).

References

1. Priestap HA, Bonafede JD, Rúveda EA (1977) *Phytochemistry* **16**: 1579.
2. Fehr C, Galindo J, Ohloff G (1981) *Helv Chim Acta* **64**: 1247.
3. Duddeck H. (1986). In: Eliel EL, Wilen SH, Allinger NL (eds.) *Top Stereochem* **16**: 293.

Exercise 3

The signals in the 1D ^{1}H NMR spectrum can be classified into two categories:

a) signals from CH fragments, that is, protons attached to bridgehead carbons or in CHBr;

b) signals from CH$_2$ fragments.

The latter can be recognized from their splitting patterns since they appear as doublets ($^2J_{HH}$ = 10 to 15 Hz) owing to the coupling with the diastereotopic geminal protons within the respective methylene groups. A further differentiation between protons of the two categories can be seen in the H,C COSY spectrum; for each methylene group there are two correlation peaks for one carbon resonance. For the signals at δ = 34.0 and 30.2 the typical artifacts can be observed exactly midway between the cross peaks of the two diastereotopic hydrogen atoms (cf. Sect. 2.5).

Long-range couplings, especially via four bonds, are represented by cross peaks only if all the atoms in the respective molecular fragments are coplanar, that is, in W arrangements.

Among the CH proton signals, that from the CHBr group (H–4^a) can be identified most easily because of its large chemical shift (δ = 4.41); the other CH signals appear at δ = 2.70, 2.50, 2.19, and 1.90. The first two with their larger δ values belong to H–1 and H–3, since these atoms are situated next to the carbonyl group (C–2). The CHBr group can be found on the other side of H–3, so it is reasonable to argue that the H–3 signal has the highest chemical shift. At this stage, this is only a hypothetical assumption, which, however, shall turn out to be correct in the following. Thus, a self-consistent signal assignment will be achieved.

In the H,H COSY spectrum the H–3 signal is connected by cross peaks with that of H–4^a and with another bridgehead hydrogen signal that can only belong to H–5 (W coupling). Thus, the signals of H–1 and H–7 can be identified as well; the H,C COSY spectrum also enables us to assign the corresponding carbons. Returning to the H,H COSY spectrum, two cross peaks can be seen correlating H–4^a with two protons in the methylene groups (δ = 1.76 and 1.68). These must be H–6^e and H–10^a, since only these atoms are in W arrangement with respect to H–4^a and, therefore, have a four-bond coupling of 2 to 3 Hz. The signal at δ = 1.76 is connected with H–3 by another cross peak; it is H–10^a, and the other (δ = 1.68) is H–6^e. A coupling also exists between H–5 and H–6^e, and there are cross peaks correlating H–7 with both H–6^e and H–10^a. This finding is a confirmation of the preceding bridgehead proton assignment. Another H–3 cross peak leads to H–10^e (δ = 2.55). This proton, being geminal to H–10^a, can also be identified by a cross peak connecting them, and the position of H–6^a (δ = 2.39) can similarly be found. Again, the position of the signals of geminal hydrogen atoms can be taken from the H,C COSY spectrum, another example of redundant information in the two COSY spectra.

The H–8 and H–9 signals remain to be identified. H–5 can be correlated by cross peaks only with the two H–9 atoms; the corresponding signals are at δ = 2.14 and 1.84 and have a common cross peak as well. Thus, the two H–8 atoms resonate at δ = 1.97 and 1.93; in the 1D spectrum their signals appear as a strongly coupling AB spin system with additional splitting due to couplings with other neighbors. The right-hand part of the AB system (δ = 1.93) is partially overlapped by the signal of H–7.

The only question still open is the stereochemical assignment. H–6^a is linked only to one H–9 by a cross peak, namely, that with δ = 1.84, which therefore represents H–9^a (W arrangement). A similar assignment of the two H–8 atoms is difficult because they have a very similar chemical shift. Here correlations to H–9^a and H–10^e are helpful: H–8^e provides the signal at $\delta = 1.97$, and H–8^a at $\delta = 1.93$.

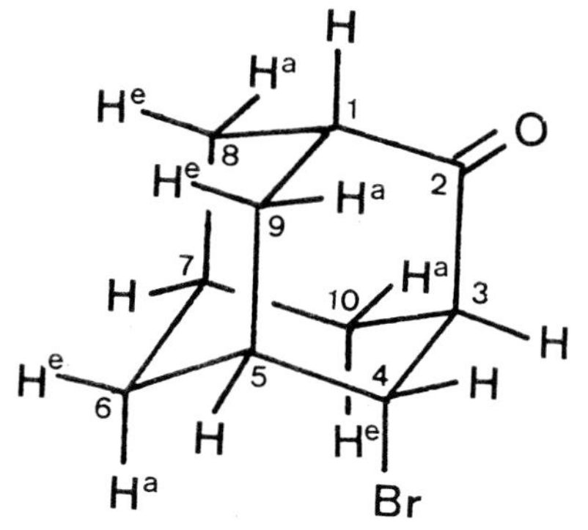

Fig. 5.3.1. Designation of the atoms in **3**.

Finally, the remaining ^{13}C resonances can be readily assigned.

The signal multiplicities of methylene protons can easily be interpreted and used for a signal assignment. In addition to the large couplings between geminal hydrogen nuclei ($^2J_{HH}$), signal splittings due to vicinal ($^3J_{HH}$) and long-range W couplings ($^4J_{HH}$) can be identified and are apparently similar in magnitude. Since each methylene proton necessarily has two vicinal bridgehead neighbors, a signal splitting into a triplet means that there is no W coupling. If it is quartet-like, there is one such long-range coupling, if it is quintet-like, there are two. Therefore, the signal at $\delta = 2.14$ must belong to H–9^e because it is the only methylene hydrogen in the molecule not having a single W partner; both positions are occupied by the bromine and the carbonyl oxygen atoms. The two H–8 can be differentiated with an analogous argument; the signal at $\delta = 1.97$ indicates only one, that at $\delta = 1.93$ two. Thus, the first is associated with H–8^e, since one W position is blocked by the carbonyl group. On the other hand, H–8^a still has both W partners, namely H–6^a and H–9^a.

The foregoing are strong arguments supporting the prior results obtained from the H,H COSY spectrum. We should not neglect to note, however, that in spite of the high magnetic field, the entire spin system is of high order, and signal splitting has to be interpreted cautiously. Such arguments belong to the category of hints and are not experimental proofs. In cases of doubt they should be given less credence than the information derived from cross peaks. Experience has shown that an evaluation of ^{1}H signal multiplicities, as described above, leads to reliable results in most cases, but reliability can deteriorate rapidly if lower magnetic fields are employed (cf. Fig. 2.1.1 and the discussion in this section).

The ^{13}C signal assignment agrees entirely with a previous one [1,2] based on empirical increment rules, the application of which is demonstrated in Exercise 4.

Signal assignment for **3**: δ_H = 2.50 (H–1), 2.70 (H–3), 4.41 (H–4), 2.19 (H–5), 2.39 (H–6a), 1.68 (H–6e), 1.90 (H–7), 1.93 (H–8a), 1.97 (H–8e), 1.84 (H–9a), 2.14 (H–9e), 1.76 (H–10a), and 2.55 (H–10e); δ_C = 45.2 (C–1), 212.5 (C–2), 54.7 (C–3), 56.6 (C–4), 34.9 (C–5), 30.2 (C–6), 26.6 (C–7), 39.3 (C–8), 35.6 (C–9), and 34.0 (C–10).

References

1. Duddeck H (1975) *Org Magn Reson* **7**: 151.
2. For the ^{13}C NMR spectroscopy of substituted adamantanones, see Duddeck H, Islam MR (1984) *Chem Ber* **117**: 554 and the references cited therein.

Exercise 4

In this example the question is to which of the three dibromoadamantanones the depicted spectra belong. The answer can be found very easily by merely inspecting the 1D ^{1}H NMR spectrum! The two CHBr peaks differ in their splittings: That at $\delta = 5.08$ is a multiplet, whereas the one at $\delta = 4.76$ is a triplet. Thus, the latter corresponds to a hydrogen atom that does not have a coupling partner in the W position (cf. Exercise 3). Checking the formulas of isomers **5** through **7**, it is apparent that there is only one such instance, namely, H–4^e in **7**. Here both W positions are blocked by the carbonyl oxygen and the bromine atom (Br–9^e):

For the assignment of the residual proton signals, we can begin with the identification of the bridgehead protons in the H,H COSY spectrum, using their correlation with H–9^a and H–4^e; H–9^a leads to H–1 and H–5, H–4^e to H–3 and H–5. This proves that H–5 is resonating at $\delta = 2.54$; H–1 and H–3 can be assigned unambiguously as well. The fourth bridgehead hydrogen (H–7) corresponds to the signal at $\delta = 2.00$ since this is the only one among the remaining peaks that does not display any doublet splitting due to geminal coupling. The signal of the methylene protons can also be easily assigned through the vicinal couplings with the bridgehead hydrogens and by inspecting the respective cross peaks in the H,C COSY spectrum. The H–6′ and H–6″ signals ($\delta = 1.87$ and 2.69, respectively), as well as those for H–8^a and H–8^e ($\delta = 1.84$ and 2.67, respectively), can be differentiated pairwise by evaluating their signal multiplicities. Please note that the peaks of H–6″ and H–8^e are partially overlapping, as are those of H–6′ and H–8^a. Finally, the stereochemical assignment of the two H–10 signals can be accomplished by identifying the cross peaks connecting H–8^e and H–10^e, as well as those connecting H–6″ and H–10^a.

Signal assignment for **7**: $\delta_H = 2.82$ (H–1), 2.97 (H–3), 4.76 (H–4), 2.54 (H–5), 1.87 (H–6′), 2.69 (H–6″), 2.00 (H–7), 1.84 (H–8^a), 2.67 (H–8^e), 5.08 (H–9), 2.12 (H–10a), and 2.16 (H–10e); $\delta_C = 54.1$ (C–1), 208.7 (C–2), 54.2 (C–3), 56.0 (C–4), 42.2 (C–5), 32.1 (C–6), 25.6 (C–7), 34.5 (C–8), 52.3 (C–9), and 41.3 (C–10).

The rearrangement reaction and the NMR data of the other isomers are reported in [1].

The ^{13}C chemical shifts can be calculated using increment rules derived from adamantanone (**8**) and 2-bromoadamantane (**9**) [1,2].

8

9

References

1. Duddeck H, Brosch D (1985) *J Org Chem* **50**: 5401.
2. Duddeck H (1975) *Org Magn Reson* **7**: 151.

Exercise 5

Both the ^1H and ^{13}C NMR spectra clearly show that the sample under investigation is a mixture. The fact that many signals appear as pairs having very similar chemical shifts and splittings but unequal intensities suggests that the mixture consists of two stereoisomers. For determining the component ratio the two signals at $\delta = 4.84$ and 4.93 are most suitable; they apparently represent the same atom in each of the two isomers, and, in addition, they do not overlap. Integration, therefore, provides reliable intensity measures. The result is that the substance is almost exactly a $1:2$ mixture.

First, the signals of the two components have to be separated by using the H,H COSY spectrum to determine the proton connectivity. The entry point is the olefinic signal of the major (minor) constituent at $\delta = 4.84$ (4.93) from which the two on the right-hand (left-hand) side of the four partially overlapping singlets at $\delta = 1.66$ and 1.64 (1.69 and 1.66) are reached. For each component and each peak the relative intensity represents three protons. The H,C COSY and DEPT spectra prove that these signals correspond to methyl groups. Since all ^1H methyl signals are singlets, the coupling to the olefinic hydrogens must be very weak; it could, for instance, be an allylic coupling. A further coupling partner of the initial olefinic proton has an ^1H chemical shift of ca. 1.05 (1.35), and from there one can go to a multiplet at $\delta = 0.80$ (~ 1.05) in the H,H COSY spectrum. Finally, this latter signal is connected by cross peaks with the two multiplets at $\delta = 3.74$ and 3.53 (3.63 and 3.59). Both belong to the same methylene group with a ^{13}C chemical shift of $\delta = 63.6$ (60.5), a typical value for a $-CH_2O-$ fragment. The signal at $\delta \approx 1.05$ (1.35) has another coupling, albeit a weak one, with the methyl signals at $\delta = 1.66$ and 1.64 (1.69 and 1.66).

The oxygen atom belongs to a hydroxy group, the proton of which affords a broadened singlet between $\delta = 1.35$ and 1.5. This signal is common to both components. The atoms identified thus far can be combined to form the following structural fragment:

$$\begin{array}{c} H_3C \\ \searrow \\ H_3C \nearrow \end{array} C = CH - \overset{|}{C}H - \overset{|}{C}H - CH_2OH$$

The ^1H NMR spectrum also reveals two additional singlets at $\delta = 1.13$ and 1.04 (1.09 and 1.02), which apparently belong to two isolated methyl groups; there are no cross peaks detectable in the H,H COSY spectrum. Evaluating the ^1H and the corresponding ^{13}C chemical shifts, we can reasonably assume that both are attached to an aliphatic quarternary carbon: Indeed, there is such a signal in the ^{13}C NMR spectrum at $\delta = 22.3$ (20.8), with an expectedly small intensity, that does not appear in the DEPT spectra. Moreover, a fragment CH_3-C-CH_3 covers the remaining three carbons and six hydrogens. The molecular formula proves the existence of two double-bond equivalents. Thus, in addition to the double bond identified above, there must be a ring in the molecule. By combining the two fragments under ring formation, we obtain chrysanthemum alcohol [1] in which are present the following two diastereomers:

10   **11**

Next, it is necessary to clarify the stereochemical assignment, that is, whether **10** or **11** is the major component in the mixture. Therefore, we have recorded an NOE difference spectrum irradiating the signal at $\delta = 0.80$, which represents H–3 of the major isomer. The experiment shows a significant signal enhancement for H–5, which is possible only in the trans-configuration (**10**). Thus, the substance is a trans/cis mixture of **10** and **11** in ratio of 2:1. In addition, the two H–1 signals of **10** can be detected in the NOE difference spectrum, as expected. Owing to the relatively large coupling constant between the H–2 and two H–1 atoms ($^3J_{HH} = 8.4$ and 6.4, respectively), the typical polarization transfer effect can be observed; that is, not all of the partial signals appear with enhanced intensity (cf. Sect. 2.3).

The stereochemical assignment of the methyl carbon atoms is based upon substituent effects on ^{13}C chemical shifts. In the isobutylene group, C–7 is in trans and C–8 in cis position with respect to C–4. Therefore, C–8 is significantly shielded as a result of the γ effect of C–4. This is why in both diastereomers the chemical shift of C–8 is 7 to 7.5 ppm smaller than that of C–7. In the cis-isomer **11** C–10 has two γ-gauche-positioned neighbors, namely, C–1 and C–5; in contrast, C–9 has none. Therefore, the C–9 and C–10 chemical shifts differ by 13 ppm (two γ gauche effects). In the trans isomer this is not clear, since each of the two methyl groups has one γ gauche neighbor (C–9 $\cdots$ C–1 and C–10 $\cdots$ C–5). This is why the two chemical shifts are very similar; no safe assignment is possible on the basis of such arguments.

It is difficult to correlate the ^{13}C and ^{1}H signals. Therefore, an expanded section of the H,C COSY spectrum is depicted in Figs. 3.5.5 and 5.5.1.

Signal assignment for **10** (trans-isomer, major component): $\delta_H = 3.74/3.53$ (two H–1), 0.80 (H–2), ca 1.05 (H–4), 4.84 (H–5), 1.66 (H–7), 1.64 (H–8), 1.13/1.04 (H–9/H–10)[1], and 1.35–1.50 (OH); $\delta_C = 63.6$ (C–1), 35.1 (C–2), 22.3 (C–3), 28.6 (C–4), 123.4 (C–5), 133.0 (C–6), 25.6 (C–7), 18.2 (C–8), and 21.3/22.6 (C–9/C–10)[1].

Signal assignment for **11** (cis-isomer, minor component): $\delta_H = 3.63/3.59$ (two H–1), ca 1.05 (H–2), 1.35 (H–4), 4.93 (H–5), 1.69 (H–7), 1.66 (H–8), 1.09 (H–9), 1.02 (H–10), and 1.35–1.50 (OH); $\delta_C = 60.5$ (C–1), 31.0 (C–2), 20.8 (C–3), 26.1 (C–4), 119.0 (C–5), 135.0 (C–6), 25.7 (C–7), 18.4 (C–8), 28.8 (C–9), and 15.4 (C–10).

---

[1] A safe and unambiguous assignment of the H–9/H–10 and C–9/C–10 signals of **10**, respectively, is not possible on the basis of the depicted spectra.

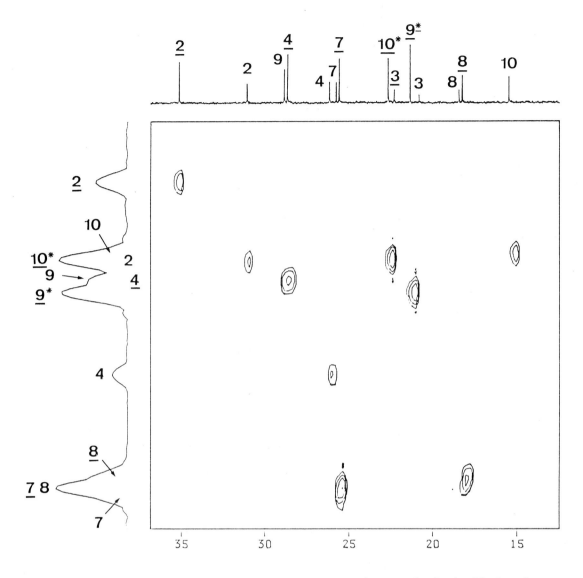

Fig. 5.5.1. Expanded section of the H,C COSY spectrum, including the assignment; the signals of the trans-isomer (**10**, major component) are underscored.

Reference

1. Banthorpe DV, Charlwood BV (1972). In: Newman AA (ed) Chemistry of Terpenes and Terpenoids. Academic Press, New York London, p 358.

Exercise 6

The relative configuration and/or conformation of compound **12** may be determined if the orientation of atoms H–6, H–6a, and H–7 is known. The signals of these proton are in the range of δ = 5.1 to 3.6, and they clearly display different splitting patterns. The assignment of the peak with the smallest chemical shift (δ = 3.68, splittings of 12.2 and 1.2 Hz) is easy; it corresponds to H–6a, since this atom is at the greatest distance from the electron-withdrawing sulfur and oxygen atoms. The C–6a atom identified in the H,C COSY spectrum has the smallest δ value (46.2). The two remaining protons, however, cannot be assigned directly because of their very similar chemical shifts; here the correlation with the respective carbon signals is needed (H,C COSY). Undoubtedly, C–6, which is directly attached to oxygen has the larger chemical shift (δ = 78.9), whereas that of C–7, carrying the much less electronegative sulfur atom, is smaller (δ = 60.6). Now, the H,C COSY spectrum proves that the proton signal at δ = 5.00, with a splitting of 12.2 Hz, belongs to H–7 and that the other at δ = 4.94 appears as a broadened singlet belonging to H–6.

This differentiation cannot be accomplished by a simple 1D selective decoupling experiment because the two signals are too close even at 400 MHz. In an earlier paper [1] it was shown that H–6 and H–7 can be assigned by means of an SPT experiment [2].

The spin system H–6/H–6a/H–7 is of type AMX, thus all coupling constants can be read from the splittings: $J(6,6a)$ = 1.2 Hz, $J(6,7)$ ≈ 0 Hz, and $J(6a,7)$ = 12.2 Hz. The latter value clearly shows that H–6a and H–7 are *antiperiplanar*; therefore, two of the four possible pairs of diastereomers can now be excluded. Each of the two others, **12C** (cis-isomer) and **12T** (trans-isomer), can adopt two different conformations through a ring inversion. These four structures are depicted in Fig. 5.6.1. The structures **12C-a,e** and **12T-a,a** can also be excluded because the torsional angle between H–6 and H–6a – 50° to 60° as estimated from Dreiding models – do not agree with their very small coupling constant found in the spectrum. A sound decision with respect to the two others, **12C-e,a** and **12T-e,e**, that is, a decision based on experimental proofs, cannot be made on the basis of the available spectral evidence. Inspecting molecular models, we find that in **12C-e,a** a torsional angle τ (H–6, H–6a) of about 90° can be achieved only if the molecule adopts a puckered half-chair conformation of the oxygen-containing six-membered ring. This means, however, that the quasi-axial phenyl ring at C–6 has to be pressed inward even more than it already is in the "relaxed" model. An analogous consideration for **12T-e,e** seems to afford a more plausible conformation. Here a torsional angle of about 90° can be easily obtained by flattening the six-membered ring. This leads to an outward motion of the phenyl ring that seems to be energetically more favored. These admittedly vague arguments suggest that the structure of **12** is that represented by formula **12T-e,e**, an assumption supported by a comparison of the ^{13}C chemical shifts of **12** and a derivative that does not bear a phenyl group at C–6 [1].

This exercise is instructive in two ways, although the information offered is not sufficient for arriving at an unambiguous solution: a) it can be seen that the assignment of only three signals among many others may be enough to give an answer to a stereochemical problem; b) it is clearly demonstrated that one single piece of information (here, the relative orientation of H–6, H–6a, and H–7) is not sufficient to provide experimental proof for the configuration and the conformation at the same time. Unfortunately, this fact is often neglected.

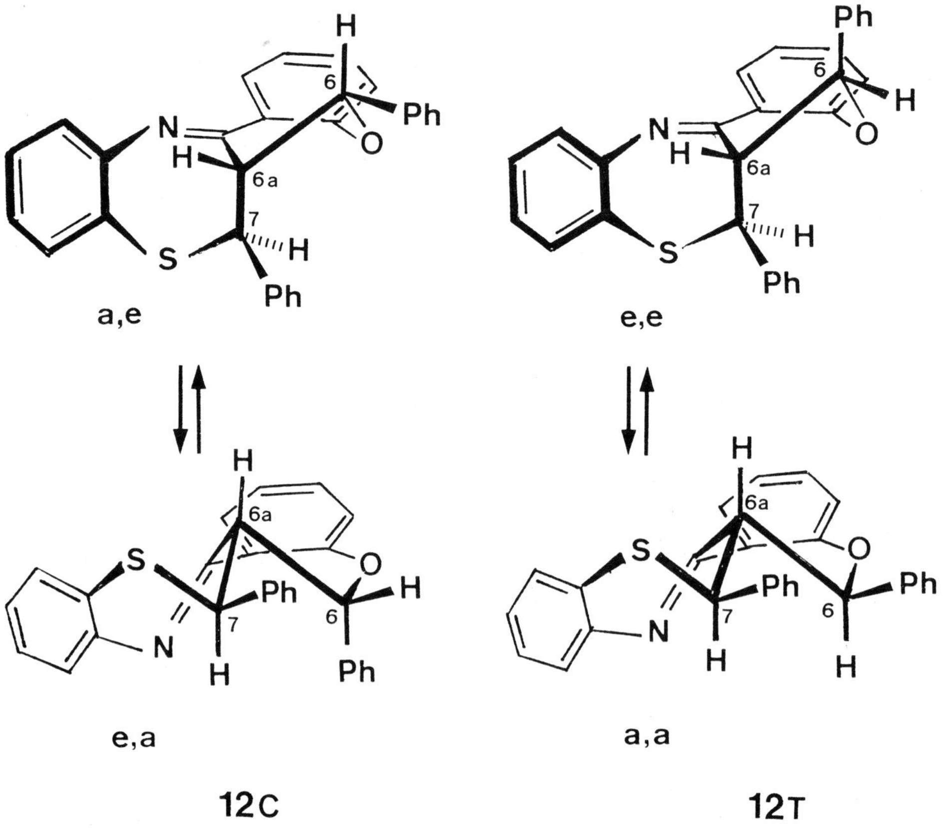

Fig. 5.6.1. Possible conformation of **12**. The designations cis and trans refer to the relative orientations of H–6 and H–6a; "a" and "e" designate the stereochemical position of the respective hydrogen atoms in the oxygen-containing ring, the first letter corresponding to H–6, the second to H–6a.

References

1. Tóth G, Szöllösy A, Lévai A, Duddeck H (1982) *Org Magn Reson* **20**: 133.
2. Martin ML, Martin GJ, Delpuech J-J (1980) Practical NMR Spectroscopy. Heiden, London.

Exercise 7

Apparently, the reaction indicates the addition of a carbene to an olefin. Thus, one does not expect to find heteroatoms in the molecule. The M^+ peak in the mass spectrum suggests the molecular formula C_8H_{12}; that is, there are three double bond equivalents. The ^1H and ^{13}C NMR spectra prove that the compound is purely aliphatic, hence the molecule contains three rings. The ^{13}C NMR spectrum displays only six signals, four for methylene groups and two for quarternary carbons. Therefore, it must contain symmetry elements other than C_1, and since the signals at $\delta = 28.8$ and 5.0 seem to represent two carbons each, there may be either a mirror plane or a C_2 axis.

It can be seen in the H,H COSY spectrum that the proton signal groups at $\delta = 2.2$ to 1.75 (intensity for six hydrogens), the singlet at $\delta = 0.75$ (two hydrogens), and the group at $\delta = 0.73$ to 0.64 (four hydrogens) are more or less isolated spin systems, that is, there is apparently no significant coupling between them. This leads to the assumption that they are separated by quarternary carbons that are spiro atoms[1].

There is a singlet at $\delta = 0.75$ in the ^1H NMR spectrum representing two hydrogens that belong to a single methylene group ($\delta = 18.6$), and they seem to be chemically equivalent. The corresponding one-bond coupling constant – obtained from the gated-decoupled spectrum – is 157 Hz. This is a relatively large value suggesting that the methylene group is incorporated in a cyclopropane ring. Since the two protons do not have significant couplings with other protons, it is reasonable to assume that the two other carbons in the cyclopropane ring are quarternary.

The ^{13}C signal at $\delta = 5.0$ representing two chemically equivalent carbons, also displays a large $^1J_{HH}$ value, namely, 161 Hz. The corresponding proton signal group is symmetrical and has the typical AA′BB′ spin-system appearance. Thus, these atoms form another cyclopropane ring connected with the first by a quarternary spiro carbon.

The remaining atoms form three methylene groups, two of which are chemically equivalent. They belong to the cyclobutane ring that was introduced during the reaction and which again is connected to the rest of the molecule via a spiro atom. The one-bond carbon-hydrogen coupling constants are 136 and 138 Hz for the carbon signals at $\delta = 28.8$ and 16.8, respectively, typical values for four-membered rings. Thus, it can be concluded that compound **13** is dispiro[2.1.3.0]octane, which was formed in the reaction of cyclobutylidene with methylenecyclopropane [1]:

---

[1] Another possibility is a propellane, which, however, would be highly strained considering the low molecular weight. The fact that the two quarternary carbons are nonequivalent is also an important argument against a propellane structure; such a small propellane without a mirror plane is hardly conceivable.

The assignment of H–5/H–7 and H–6 can be determined from the H,C COSY spectrum. Finally, the two quarternary carbons should be differentiated. In the hetero-NOE difference experiment (Fig. 3.7.2c) the protons H–5/H–7 have been irradiated, but only one of the two quarternary carbons responds, namely, that with $\delta = 24.8$; it is C–4.

Signal assignment for **13**: $\delta_H = 0.7$ and 0.67 (two H–1 and two H–2), 2.13, and 1.96 (two H–5 and two H–7), 2.03 and 1.79 (two H–6), and 0.75 (two H–8); $\delta_C = 5.0$ (C–1/C–2), 19.2 (C–3), 24.8 (C–4), 28.8 (C–5/C–7), 16.8 (C–6), and 18.6 (C–8).

Reference

1. Brinker UH, Weber J (1986) *Tetrahedron Lett* **27**: 5371.

Exercise 8

The molecular formula of **14** proves the existence of four double-bond equivalents: one corresponds to a double bond (olefinic proton with $\delta = 5.42$, sp^2–C with $\delta = 169.9$, and sp^2–CH with $\delta = 120.8$), and the other to a carbonyl group ($\delta = 203.3$). Thus the compound contains a C=CH and a C=O group, and two rings.

Four separate signals can be identified in the ^{1}H NMR spectrum ($\delta = 2.53, 2.33, 2.16,$ and 1.77). As can be seen in the H,H COSY plot, each of the three protons with the larger chemical shifts couples with the two others; the fourth with $\delta = 1.77$ couples only with that having the largest δ value ($\delta = 2.53$). There are no further recognizable cross peaks for these four ^{1}H signals, so it is reasonable to assume that the respective fragments are isolated from the rest of the molecule by quarternary carbons. According to the H,C COSY and DEPT spectra, the two protons with $\delta = 2.53$ and 1.77 belong to the same methylene group ($\delta_C = 40.5$) that neighbors two methine groups ($\delta_H = 2.33$, $\delta_C = 57.3$ and $\delta_H = 2.16$, $\delta_C = 49.4$). Two alternative fragments can be constructed:

>CH—CH$_2$—CH<   or   >CH — CH — CH$_2$—
                                                        |

A   **B**

The ^{1}H spin system formed by these three protons is interesting in two aspects:

a) The two methine signals have very similar splitting patterns; therefore, the coupling constants involved must be similar as well. This finding is unfavorable to structure **B**.

b) For each of the two methine protons the two vicinal couplings with the methylene protons are significantly different. The coupling constants with the proton resonating at $\delta = 1.77$ are very small, indicating that the torsional angle is about 90°. As already stated, the compound is bicyclic. Assuming that fragment **A** is correct, there are two possibilities regarding its incorporation into a bicyclic ring system; it could be either a bornane (**C**) or a pinane (**D**) skeleton:

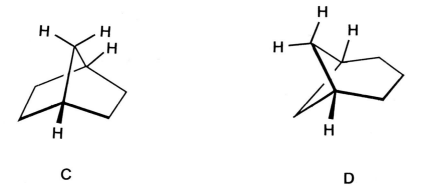

C   **D**

As noted, the protons in the CH$_2$ bridge couple differently with the methine protons. This strongly favors the less symmetric alternative, that is, the pinane structure **D**.

The methyl signal at $\delta = 1.74$ ($J = 1.2$ Hz), split into a doublet, shows a common cross peak with the signal of the olefinic proton (allylic coupling). The olefinic carbon resonances ($\delta = 169.9$ and 120.8) are far from each other; the double bond is apparently conjugated to the carbonyl group. This suggests an unsaturated fragment: $O=C-CH=C-CH_3$.

The residual atoms are a quarternary carbon with $\delta = 53.6$ and two methyl groups with chemical shifts of $\delta = 26.3$ and 21.8 for carbon, and $\delta = 1.22$ and 0.72 for hydrogen. Both are expected to be geminal; therefore, a third structure fragment, CH_3-C-CH_3, is identified. Putting all substructures together, we obtain the formula for verbenone (**14**); the configuration for its levorotatory form is depicted in Fig. 5.8.1.

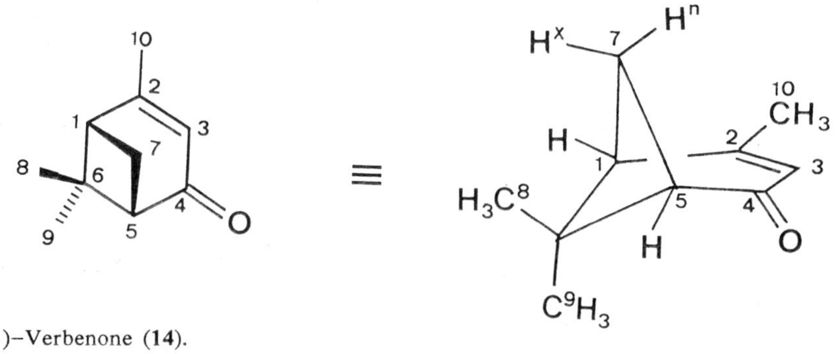

Fig. 5.8.1. $(-)$-Verbenone (**14**).

Finally, the signal assignment for the hydrogen atoms in the fragment C–1/C–7/C–5 and in the two geminal methyl groups has to be ascertained. This can be achieved by evaluating the NOE difference spectrum. Irradiation of the methyl protons H–10 (Fig. 3.8.2b) affords a significant signal enhancement only for the CH proton with $\delta = 2.16$, so this should be H–1. If the irradiation is found at the location of the methyl signal, $\delta = 1.22$ (Fig. 3.8.2c), a clear NOE is visible for one of the H –7 atoms ($\delta = 2.53$). Therefore, this is the exo-positioned H–7[x], and the methyl signal irradiated belongs to H–8[1]. As expected, small signal enhancements are also observed for the two bridgehead protons and for H–9. The counterexperiment, irradiation at H–9 (Fig. 3.8.2d), gives an NOE at H–8 only and a weak one for the olefinic H–3.

The endo-positioned H–7[n] ($\delta = 1.77$)[1] does not have a significant coupling with the two bridgehead atoms H–1 and H–5 (cf. H,H COSY spectrum) because the respective torsional angles H–7[n]/C–7/C–1/H–1 and H–7[n]/C –7/C–5/H–5 are close to 90°. This can clearly be seen in Dreiding models.

At last it should be mentioned that two weak allylic couplings exist between the olefinic H–3 and the two bridgehead protons; they can be identified by small but significant cross peaks in the H,H COSY plot.

Signal assignment for **14**: $\delta_H = 2.16$ (H–1), 5.42 (H–3), 2.33 (H–5), 1.77 (H–7[n])[1], 2.53 (H–7[x])[1], 0.72 (H–8), 1.22 (H–9), and 1.74 (H–10); $\delta_C = 49.4$ (C–1), 120.8 (C–2), 169.9 (C–3), 203.3 (C–4), 57.3 (C–5), 53.6 (C–6), 40.5 (C–7), 21.8 (C–8), 26.3 (C–9), and 23.3 (C–10).

---

[1] Notation: Superscript "x" stands for exo and "n" for endo.

Exercise 9

A number of important ^{13}C signals can be identified by a comparison with the data for many related steroids, which has been compiled in a review [1]. The signal at $\delta = 79.5$ belongs to C–3, that at $\delta = 61.5$ to the methylene carbon of the ethoxy group (C–23), and that at $\delta = 55.5$ to the methoxy group attached to C–3. The corresponding proton signals are H–3 with $\delta = 3.10$, the two diastereotopic H–23 atoms with $\delta = 4.01$ and 3.75, and the methoxy protons with $\delta = 3.32$. Next, the spin system consisting of H–15α, H–15β, H–16, H–17, and H–20 has to be identified. The initial hypothesis is that the signal of H–16 should resonate at a relatively high frequency (large δ value) since it is in geminal position with respect to a sulfur atom. In addition, there should be two cross peaks connecting it with the two geminal H–15 atoms in the H,H COSY plot. This is the case for the double doublet appearing at $\delta = 3.93$; the fact that the two H–15 atoms with $\delta = 2.70$ and 2.31 are indeed geminal is proved by the H,C COSY and the DEPT spectra.

An AX-type spin system consisting of two doublets at $\delta = 3.79$ and 2.44 can be recognized. These signals can only represent the remaining H–20 and H–17 atoms; all other protons in the molecule have more than one coupling partner. The signal at $\delta = 3.79$ should be assigned to H–20 because of the two neighboring carbonyl functionalities. On the other hand, there is a weak cross peak indicating a coupling between H–16 and H–17, which, however, is apparently of very low magnitude, since both signals do not display a corresponding splitting but only a line broadening. All proton signals are far enough away from each other so that coupling constants can be read from their splittings: $J(17,20) = 4.4$ Hz, $J(15\alpha,15\beta) = 15.5$ Hz, and the two $J(15,16) = 5.8$ and 9.7 Hz. It is possible, but a little risky, to derive the stereochemistry of this part of the molecule, as shown in Fig. 5.9.1, from these coupling constants. NOE experiments provide much more reliable results.

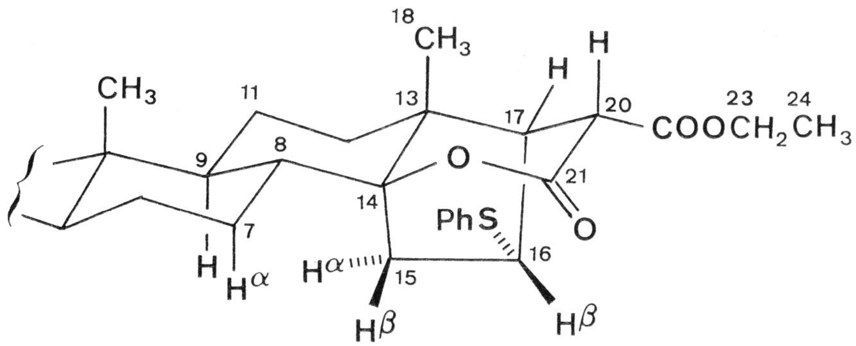

Fig. 5.9.1. Stereoprojection of 15.

Irradiation of the signal at $\delta = 3.79$ (H–20, Fig. 3.9.5c) provokes signal enhancements at H–17 and H–18, as well as at one of the H–23 signals – apparently because the other H–23 resonance is very close to the irradiation position and may be affected. The spatial proximity of H–18 and H–20 proves that H–20 is in the axial position with respect to the lactone ring. Irradiation of the H–16 signal (Fig.

3.9.5b) gives rise to NOEs at H−17, at the two ortho protons of the phenyl ring ($\delta = 7.38$), and at only one of the two H−15 atoms ($\delta = 2.70$) which is the one with the larger vicinal coupling constant (9.7 Hz). This finding and the fact that the coupling constant $J(16,17)$ must be close to zero prove that H−16 and NOE-affected H−15 are in β position. Only in such case are the torsional angles τ(H−16$\cdots$ H−17) $\approx 90^{\circ}$ and τ(H−16 $\cdots$ H−15β) $\approx 0^{\circ}$ in agreement with these coupling constants. This also holds for the coupling $J(15\alpha,16)$ with $\tau \approx 120^{\circ}$.

There are further hints that this assignment is correct. Irradiation of H−15α (Fig. 3.9.5f) leads to an enhancement of the signal at $\delta = 1.04$. The appearance of this peak (triplet with further fine splittings) is what one would expect for H−9, which is close to H−15α in space and has three vicinal coupling partners, two antiperiplanar (H−8 and H−11β) and one gauche oriented (H−11α).

On the other hand, a slight but significant NOE enhancement is found for a signal at $\delta = 0.96$ if H−15β is irradiated (Fig. 3.9.5d). This can only be the nearby H−7α, and its signal appears as a broadened quartet because of the existence of three strong couplings with H−6β, H−7β, and H−8, as well as a weaker one with H−6α.

A detailed analysis of the H,H and H,C COSY spectra and a comparison of the ^{13}C signal with those of similar steroidal compounds [1] allow the assignment of many more atoms in the A, B, and C rings. This, however, is not essential for answering the initial question.

Reference

1. Blunt JW, Stothers JB (1977) *Org Magn Reson* **9**: 439.

Exercise 10

This exercise should begin with the ^{1}H and ^{13}C signal assignments. The two signals at $\delta = 6.57$ and 5.86 ($J = 9.3$ Hz) belong to the double bond; that with the smaller chemical shift is in α position and that with the larger is in β position with respect to the carbonyl group. The corresponding carbons are α–C, with $\delta = 125.6$, and β–C, with $\delta = 151.7$. The values and their sequences are in good agreement with well-known data from α,β–conjugated enones [1].

The two carbonyl signals (cf. the horizontal trace of the COLOC spectrum) are easy to differentiate: That at $\delta = 201.6$ belongs to the ketone and that at $\delta = 178.2$ to the lactone C–12.

The proton signal at $\delta = 4.13$ obviously refers to H–6, since C–6 bears an oxygen atom. It is split into a double doublet as a result of two couplings of approximately 10 to 12 Hz; apparently H–6 has two antiperiplanar neighbors, namely, H–5 and H–7. This proves that the two six-membered rings are trans fused. H–6 is on the β side; H–5 ($\delta = 2.40$) and H–7 ($\delta = 1.65$) can be identified readily in the H,H COSY plot and are α positioned. In addition, the H–11 atom ($\delta = 2.32$) can easily be recognized by its splitting into a double quartet.

Among the methyl signals only that of H–13 ($\delta = 1.22$) can be identified directly by its doublet structure. It is reasonable to assume that the methyl singlet with the higher chemical shift ($\delta = 1.52$) belongs to H–14, since this group is in geminal position with respect to the hydroxy group ($\delta = 2.85$).

The H,H COSY plot leads from the H–7 signal to both H–8 signals ($\delta = 1.95$ and 1.45) and from there to the two H–9 signals ($\delta = 2.00$ and 1.54). It is possible to discriminate between the H–8 and H–9 signals by inspecting the H,C COSY spectrum, which displays the signals of geminal disstereotopic protons in vertical traces. Owing to the small distances between the H–7, H–8, and H–9 resonances, their signals are of high order, and a simple evaluation of the splittings in terms of coupling constants is not possible. The two peaks at $\delta = 2.00$ and 1.95, however, have a doublet-like shape. Thus, it can be assumed that they are associated with the equatorial H–9β and H–8, respectively, because each of these nuclei possesses only one partner with a large coupling constant, namely, the respective geminal proton. This also corresponds to the general experience that equatorial hydrogens are more deshielded than axial ones.

The ^{13}C signal assignment for the aliphatic carbons can be derived from the H,C COSY spectrum; their chemical shift values are collected below. According to the preceding assumption concerning the ^{1}H chemical sequence of H–14 and H–15, it emerges that C–14 resonates at $\delta = 23.7$ and C–15 at $\delta = 19.7$. This is indeed correct, as proved by some COLOC peaks correlating C–4 and H–14, as well as C–10 and H–15, pairwise.

The NOE difference spectra provide information about the stereochemistry of the sesquiterpene lactone. If the signal of the β-positioned H–6 is irradiated (Fig. 3.10.5c), significant intensity enhancements are found for H–11, H–14, and H–15, as well as a weaker one for the H–8 atom with the small δ value (1.45). This experiment affords several proofs: Both methyl groups 14 and 15 are β–oriented and therefore axial; both six–membered rings are transfused (see above); H–11 is also in β position – that is, methyl group 13 is quasi-equatorial (α); the two H–8 nuclei can be discerned with respect to their stereochemical position – the one affected by the NOE is axial (β), which is in agreement with the previous evaluation of its signal shape. Another NOE experiment (irradiation at H–15,

Fig. 3.10.5d) allows identification of the position of the two H−9 nuclei because the one that suffers from the NOE ($\delta = 2.00$) is the equatorial H−9β. The irradiation of H−15 causes the nearby H−13 signal to be affected by a spillover, and this is why a weak signal enhancement for H−11 is found. If the signal of the olefinic hydrogen in β position with respect to the carbonyl group is irradiated, a small NOE can be observed for H−15. This is a hint — although not a very convincing one — that compound **16** has structure **17**.

Much clearer is the proof for the orientation of the enone moiety in the molecule (**17** or **18**), which is present in the COLOC spectrum because pairwise correlations exist between the carbonyl carbon (C−1) and H−15, between C−3 and H− 14, and even between C−3 and the hydroxy proton.

Thus, compound **16** is vulgarin [2−6], which is also known as tauremisin.

Signal assignment for **16**: δ_H = 5.86 (H−2), 6.57 (H−3), 2.40 (H−5), 4.13 (H−6), 1.65 (H−7), 1.95 (H−8α), 1.45 (H−8β), 1.54 (H−9α), 2.00 (H−9β), 2.32 (H−11), 1.22 (H−13), 1.52 (H−14), 1.18 (H−15), and 2.85 (OH); δ_C = 201.6 (C−1), 125.6 (C−2), 151.7 (C−3), 70.0 (C−4), 54.5 (C−5), 79.5 (C−6), 52.3 (C−7), 22.6 (C−8), 34.2 (C−9), 48.2 (C−10), 40.5 (C−11), 178.2 (C−12), 12.4 (C−13), 23.7 (C−14), and 19.7 (C−15).

References

1. Pretsch E, Clerc T, Seibl J, Simon W (1983) Tables of Spectral Data for Structure Determination of Organic Compounds. Springer, Berlin Heidelberg.
2. Honwad VK, Siscovic E, Rao AS (1967) *Tetrahedron* **23**: 1273.
3. Geissman TA, Lee K−H (1971) *Phytochemistry* **10**: 663:
4. Gonzalez-Gonzalez A, Bermejo-Barrera J, Breton-Funes JL, Fajardo M (1973) *An Quim* **69**: 667.
5. Ando M, Tajima K, Takase K (1979) *Bull Chem Soc Jpn* **52**: 2737.
6. Metwally MA, Jakupovic J, Youns M I, Bohlmann F (1985) *Phytochemistry* **24**: 1103.

Exercise 11

The molecule contains two olefinic protons ($\delta = 4.92$ and 4.67) that have a small common coupling and belong to a methylene group (H,C COSY and DEPT spectra); the carbon has a chemical shift of $\delta = 112.7$. Both protons are coupled to a methyl group (^{1}H: $\delta = 1.70$, ^{13}C: $\delta = 22.4$), the coupling constant being rather small (possibly an allylic coupling). Since the only remaining olefinic carbon is quarternary ($\delta = 140.6$), the molecular fragment is an isopropylidene group. The methyl protons there have a further coupling with the hydrogen resonating at $\delta = 2.78$ that belongs to a methine fragment ($\delta_C = 45.7$) directly attached to the isopropylidene group. Cross peaks in the H,H COSY plot for this latter proton indicate the presence of a neighboring methylene group ($\delta_H = 3.39$ and 3.19, $\delta_C = 45.6$) on one side, but no other proton coupling partner thereafter. On the other side of the methine group is another methine ($\delta_H = 2.90$, $\delta_C = 40.1$), which again is in a branching position. It is attached to a third methine fragment ($\delta_H = 3.73$, $\delta_C = 65.0$) and to a methylene ($\delta_H = 2.29$ and 2.13, $\delta_C = 33.3$). Thus, fragment **A** is created:

According to the molecular formula there are two remaining carbons ($\delta = 172.7$ and 169.3), three hydrogens (a very broad signal at $\delta = 5$ to 7, indicating the presence of three acidic protons), one nitrogen, and four oxygens. These atoms can be combined to form two carboxylic acid functionalities and one NH group. In addition, the formula proves the existence of four double bond equivalents, three of which are covered by the isopropylidene and the two carboxyl groups. Thus, compound **19** is monocyclic and, apparently, NH is a bridging group. Such a connection via the two terminal methylene groups of fragment **A** is excluded because the resulting molecule would not be an α-amino acid. Therefore, the ring has to be closed using one methylene and the methine group. According to which methylene is used, an azetidine (four-membered ring) or a pyrrolidin derivative (five-membered ring) emerges. The latter alternative is the more likely one because the molecule is a proline derivative.

The NOE difference experiment depicted in Fig. 3.11.4d (irradiation of the signal at $\delta = 3.73$) affords a weak but significant NOE for a proton ($\delta = 3.19$) of the methylene group on the right-hand side of fragment **A**. So this and the irradiated proton are positioned on the same side of the ring and are relatively near each other, a situation possible only in the proline derivative **B**.

It is very probable that the compound is an L-amino acid, and so, under this assumption, we can fix the C–2 configuration. The C–3 configuration can be derived from the NOE difference spectra in Figs.

3.11.4e and f. When one of the H–6 is irradiated, a significant signal enhancement is found for H–2, whereas this effect is much less if the other H–6 is irradiated. This observation is possible only if H–2 and C–6 are cis oriented. The counterexperiment (irradiation of H–2 and NOEs for the two H–6 atoms, Fig. 3.11.4d) confirms this interpretation. Irradiation of the olefinic proton with $\delta = 4.67$ (Fig. 3.11.4 c) provokes an intensity enhancement for that H–5 which is cis positioned with respect to H–2. This shows that the isopropylidene group is on the same side of the ring, and the configuration of the molecule is established: compound **19** is kainic acid [1].

The two H–9 signals are easily differentiated by inspecting the NOE difference spectrum in Fig. 3.1.4g. The H–9 atom absorbing at higher frequency and methyl group 10 are on the same side of the double bond. Even the two diastereotopic H–6 atoms can be discriminated: H–2 is close to H–6' (Fig. 3.11.4d) and H–10 to H–6" (Fig. 3.11.4g).

Signal assignment for **19**: δ_H = 3.73 (H–2), 2.90 (H–3), 2.78 (H–4), 3.39 (H–5α), 3.19 (H–5β), 2.29 (H–6'), 2.13 (H–6"), 4.92 (H–9a), 4.67 (H–9b), 1.70 (H–10), and 5–7 (three acidic protons); δ_C = 172.7/169.3 (C–1/C–7), 65.0 (C–2), 40.1 (C–3), 45.7 (C–4), 45.6 (C–5), 33.3 (C–6), 140.6 (C–8), 112.7 (C–9), 22.4 (C–10).

Reference

1. Beilstein, Handbook of Organic Chemistry vol 22 III/IV suppl. series. Springer, Berlin, Heidelberg, p 1523.

Exercise 12

The IR band clearly shows that the molecule contains a carboxylic acid function and a hydroxy group. Since the molecule is highly unsaturated, the latter functionality may be a phenol, and this is confirmed by the signal position ($\delta = 8.06$). Assignment of the ^{13}C signal of the carbonyl group is not obvious, since the largest chemical shift is $\delta = 162.9$, a value that is rather small; it has to be therefore concluded, that **20** is an α,β–unsaturated ester or lactone. In the ^1H NMR spectrum an AX spin system with $\delta = 7.65$ and 6.70 and $J = 9.4$ Hz can be found, and the corresponding directly attached carbons resonate at $\delta = 144.7$ and 111.7, respectively. The great differences within the pairs of these chemical shifts, as well as the δ value of the carbonyl carbon, are very characteristic for coumarin derivatives.

If our assumption of coumarin is correct, the high-frequency signal ($\delta = 7.65$) of the AX system is H–4 and that of lower frequency H–3. Both have very small cross peaks connecting them with the two singlets at $\delta = 7.17$ and 7.04. Considering the splittings of these four proton peaks, the couplings – except that between H–3 and H–4, of course – must be so weak that they hardly lead to line broadenings. This is consistent with only one molecular arrangement, namely, that of a coumarin that is 6,7– disubstituted; that is, the two protons in the benzoic ring are para positioned with respect to each other. One of these, $\delta = 7.17$, has another small but significant cross peak indicating a coupling with two aliphatic protons, $\delta = 3.36$; they both belong to a methylene group with a ^{13}C chemical shift, $\delta = 27.9$. The proton signal is split into a doublet ($\delta = 3.36$, $J = 7.2$ Hz), and this coupling leads to a triplet-shaped signal of a vicinal olefinic hydrogen with $\delta = 5.31$. The corresponding double bond is apparently trisubstituted, since one cannot find a vicinal olefinic coupling partner. The only significant cross peak leads to the methyl signal with the largest ^1H chemical shift ($\delta = 1.69$). There is another correlation signal, however, which is so extremely weak that it is unwise to use it as an argument for connectivities; thus, one position at the double bond is still unoccupied. The remaining two methyl signals ($\delta = 1.66$ and 1.57) are weakly coupled to an olefinic proton at $\delta = 5.08$, the cross peaks of which lead into the left part of the signal between $\delta = 2.2$ to 2.0. This signal corresponds to four protons that are situated in two methylene groups (cf. the DEPT spectrum). In the H,H COSY plot this diagonal peak is very broad; it is reasonable to assume that it overlaps the cross peaks. In other words, the two methylene groups are neighbors and form an ethylene fragment. The other end of the ethylene group should occupy the fourth position in the former double bond. Combining these fragments results in a side chain containing ten carbon atoms and two double bonds:

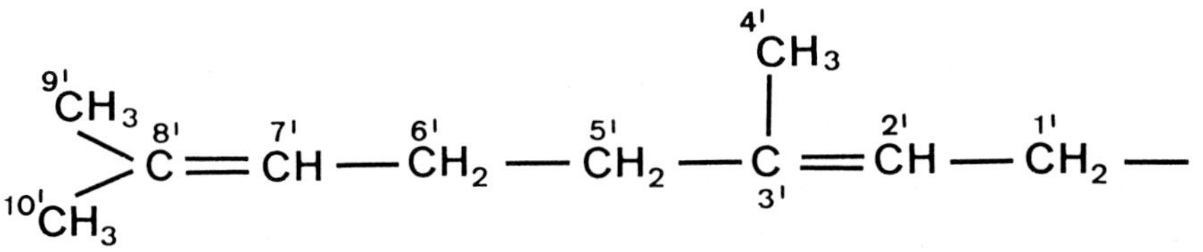

The C_{10} fragment — the geranyl substituent — is famous in natural products chemistry. Thus, the stereochemistry of the double bond C-2'/C-3' should be E, and this can easily be proved by an NOE difference experiment: When H-4' is irradiated, the signal of H-1' is strongly enhanced; this would not happen in the Z configuration.

In light of the ^{1}H,^{1}H connectivity, the very weak cross peaks in the H,H COSY spectrum (H-1' to H-5', H-2' to H-5') make sense. All ^{13}C signals of the protonated carbons can be assigned readily using the H,C COSY spectrum.

The two methyl groups 9' and 10' can be differentiated by their ^{13}C chemical shifts. The one with the smaller value ($\delta = 17.6$) corresponds to C-9' because C-6' is in the syn position and exerts a γ gauche effect on C-9'. This situation does not hold for C-10', so this nucleus is not shielded and resonates at $\delta = 25.7$.

Two points remain to be discussed: (a) which of the two substituents, geranyl and hydroxy, is in the 6 and which is in the 7 position? This is related to the question of which of the two signals at $\delta = 7.17$ and 7.04 is H-5 and which is H- 7. (b) Can the signals of the quarternary carbons be assigned?

The COLOC spectrum answers both questions: The two ^{13}C signals at $\delta = 158.8$ and 153.9 correspond to oxygenated carbons; one is C-8a, and the other carries the hydroxy group. For both there are COLOC peaks indicating long-range couplings with the two H-5/H-8 protons; these peaks cannot answer our question. For the carbon with the smaller δ value ($\delta = 153.9$, however, there is a peak for a coupling with H-4, proving that it corresponds to C-8a. This assignment is confirmed by the existence of a peak for a coupling between the C-OH carbon and H-1'. The H-5 proton can be identified by the cross peak for C-4/H-5, and this simultaneously allows the C-5 and H-8/C-8 assignment. Now we can see that C-1' has long-range couplings with H-5, and this is possible only if the geranyl side chain is attached to C-6. The remaining quarternary carbons C-4a, C-3', and C-8' can be assigned by couplings C-4a/H-3, C-3'/H-1', and C-3'/H-4', as well as by C- 8'/H-10' and C-8'/H-9'. All other COLOC peaks represent reasonable couplings too.

It is interesting to note that many small signal enhancements can be seen in the NOE difference spectrum for nuclei many bonds apart from the irradiation position (H-4'). The reason is the flexibility of the side chain, which can adopt a variety of conformations. Even the protons H-3 and H-4 are affected.

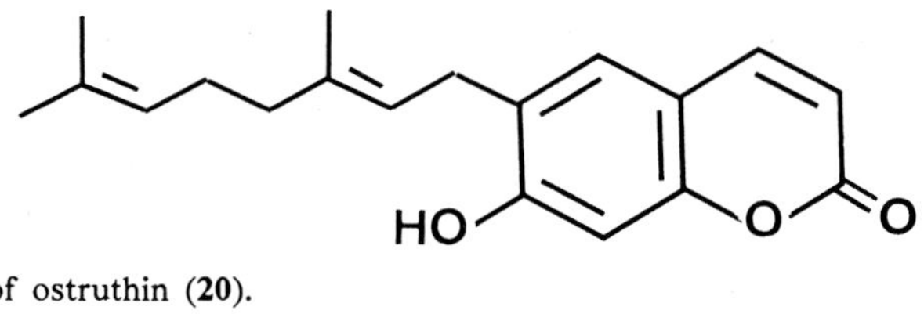

Fig. 5.12.1. Structure of ostruthin (**20**).

Thus, **20** turns out to be 7-hydroxy-6-geranylcoumarin, a compound known in the literature as ostruthin [1]. ^{13}C NMR data have been published [2,3].

Signal assignment for **20**: δ_H = 6.70 (H–3), 7.65 (H–4), 7.17 (H–5), 7.04 (H–8), 3.36 (H–1'), 5.31 (H–2'), 1.69 (H–4'), 2.10–2.05 (H–5'), 2.15–2.10 (H–6'), 5.08 (H–7'), 1.57 (H–9'), 1.66 (H–10'), and 8.16 (OH); δ_C = 162.9 (C–2), 111.7 (C–3), 144.7 (C–4), 112.0 (C–4a), 128.1 (C–5), 126.2 (C–6), 158.8 (C–7), 103.0 (C–8), 153.9 (C–8a), 27.9 (C–1'), 120.9 (C–2'), 138.1 (C–3'), 16.1 (C–4'), 39.6 (C–5'), 26.4 (C–6'), 124.0 (C–7'), 131.6 (C–8'), 17.6 (C–9'), and 25.7 (C–10').

References

1. Murray RDH (1975) Naturally occurring plant coumarins. In: Hertz W, Grisebach H, Kirby GW (eds) Fortschritte der Chemie Organischer Naturstoffe. Springer, Wien, vol 35, p 199
2. Patra A, Mitra AK (1981) *Org Magn Reson* 17: 222.
3. Duddeck H, Kaiser M (1982) *Org Magn Reson* 20: 55.

Exercise 13

First, the respective ^{1}H and ^{13}C signals for the monosaccharide and aglycone are separated. For the sugar the best entry point is the anomeric carbon ($\delta = 97.9$), to which the proton signal at $\delta = 5.97$ (doublet, 2.5 Hz) can be assigned using the H,C COSY plot. H–1 has two coupling partners at $\delta = 2.33$ and 2.22, which are connected to each other by a large coupling constant of 13.3 Hz and attached to the same carbon, namely, C–2 ($\delta = 29.4$). Thus, the saccharide is a 2-desoxy sugar. The two H–2 cross peaks in the H,H COSY spectrum lead to H–3 ($\delta = 5.40$, multiplet). The signal of H–4 ($\delta = 5.45$) is very near that of H–3, so the cross peak associated with $J(3,4)$ is very close to the diagonal and difficult to detect. Nevertheless, it is not possible to confuse H–3 and H–4 even if the H,H COSY plot is ambiguous: Since the H–3 signal must be a multiplet due to the coupling with the two H–2 signals, the signal at $\delta = 5.45$ cannot correspond to H–3. The cross peak connecting H–3 and H–4 is weak because the coupling constant $J(4,5)$ is apparently very small; the H–4 signal appears as a broadened singlet. Finally, cross peaks lead from H–5 ($\delta = 4.22$) to two H–6s with $\delta = 4.06$ and 4.04.

The coupling constants between H–1 and the two H–2 atoms are relatively small, proving that none of the H–2s is in antiperiplanar orientation with respect to H–1. Thus, H–1 is in the equatorial and the aglycone in the axial position. The signal of H–3, however, contains a 9 to 10 Hz splitting, indicating that this hydrogen is axial. The axial H–2, being antiperiplanar to H–3, resonates at $\delta = 2.33$ (H–2β). The other H–2 (H–2α) is equatorial and has a smaller coupling with H–3. This significant difference in the $J(2,3)$ values is a strong indication that the molecule contains a rigid six-membered ring, rather than a flexible, pseudorotating five-membered one; that is, it is a pyranoside. This is confirmed by the fact that vicinal coupling constants with magnitudes of about 10 Hz are found; such values only appear for fixed antiperiplanar proton orientations. Apparently, the couplings $J(3,4)$, as well as $J(4,5)$, are relatively small. The position of H–5 can be assumed to be axial because the oxymethylene group containing C–6 will invariably adopt the equatorial position.

Since the substance is readily soluble in chloroform, and in view of the existence of three ^{13}C signals around $\delta = 170$, it is reasonable to assume that the saccharide part of the molecule is peracetylated.

The molecular formula, $C_6H_3N_2O_5$, applies to the rest of the molecule, that is, to the aglycone. All six carbon signals are within the typical olefinic/aromatic range: Three carbons are quarternary and three belong to CH groups. Thus, the aglycone seems to be a dinitrophenyloxy residue. The substitution pattern can be recognized easily from the splittings of the ^{1}H signals: one signal appears at $\delta = 8.72$ as a doublet ($J = 2.7$ Hz), one at $\delta = 8.39$ as a double doublet ($J = 2.7$ and 9.3 Hz), and the third at $\delta = 7.52$, again as a doublet ($J = 9.3$ Hz). This is consistent with a 2',4'-dinitrophenoxy group.

All ^{1}H and ^{13}C chemical shifts can be verified by calculations using increment rules [1]. The ^{13}C chemical shift difference between C–2' and C–4' is too small for a safe assignment by this method; their signals may be interchanged. A heteronuclear NOE experiment irradiating H–5' would be helpful. Since, however, the structure of 21 is already elucidated, we can waive such measurement.

Thus, compound 21 is 2',4'-dinitrophenyl-2-desoxy-α-D-galactopyranoside [2].

Signal assignment for **21**: δ_H = 5.97 (H-1), 2.22 (H-2α), 2.33 (H-2β), 5.40 (H-3), 5.45 (H-4), 4.22 (H-5), 4.06 and 4.04 (two H-6), 8.72 (H-3'), 8.39 (H-5'), 7.52 (H-6'), 2.15, 1.99, and 1.92 (three acetyl-CH$_3$); δ_C = 97.9 (C-1), 29.4 (C-2), 65.2 (C-3), 65.9 (C-4), 69.1 (C-5), 61.8 (C-6), 153.8 (C-1'), 141.3 and 139.7 (C-2'/C-4', assignment uncertain), 121.7 (C-3'), 128.6 (C-5'), 117.4 (C-6'), 170.1, 169.9, and 169.3 (three acetyl-C=O), 20.7, 20.6, and 20.5 (three acetyl-CH$_3$).

References

1. Pretsch E, Clerc T, Seibl J, Simon W (1983) Tables of Spectral Data for Structure Determination of Organic Compounds. Springer Berlin Heidelberg New York.
2. Bielawska H, Michalska M (1986) *J Carbohydr Chem* **5**: 445.

Exercise 14

Since all five-membered rings are cis fused, there are four conceivable configurations, differing in the relative orientation of the pairs of bridgehead hydrogen atoms H–3/H–4 and H–10/H–11. If the two pairs are syn configurated, structure A is obtained; if they are anti structured, **B** results. Both molecules possess a C_2 rotation axis that is peripendicular to and that bisects the bond between C–3 and C–10[1]. There are no symmetry elements – except C_1, of course – in the isomers having a syn-anti or anti-syn combination. These can be excluded from further discussion because the ^{13}C NMR spectrum proves that compound **22** has a symmetry element, thus making carbons equivalent pairwise.

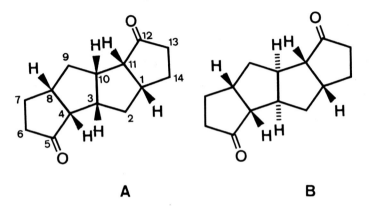

A   **B**

Fig. 5.14.1. Symmetrical configurations of **22** .

The ^{13}C resonances can be divided into two groups: Three signals appearing at $\delta = 52.5, 49.1$ and 43.8 belong to methine and the three others to methylene groups. The latter have different combinations of neighbor groupings. C–2/C– 9 is situated between two bridgehead fragments, C–14/C–7 is attached to a methine and a methylene, and C–13/C–6 has only one protonated neighbor, namely, the methylene group C–14/C–7[2]. These differences provide a means of identifying the methylene protons in the H,H COSY spectrum. The two protons at $\delta = 1.80$ and 0.75 are the two H–2/H–9 protons, since cross peaks can be found connecting them with the methine protons with $\delta = 2.70$ and 2.58. Analogously, it is possible to show that the signals of H–7/H–14 are at $\delta = 1.77$ and 1.61 and those of H–6/H–13 at $\delta = 2.07$ and 2.05. Now, the methine protons can be identified as well: That with the largest chemical shift ($\delta = 2.70$) couples with both H–2/H–9s, as well as with both H–14/H–7s, that is, it is H–1/H–8. The next ($\delta = 2.58$) couples only with the two H–2/H–9s, thus it is H–3/H–10. For this signal one can find a cross peak leading to the third methine signal belonging to H–11/H–4 ($\delta = 2.42$). Probably, this is not a four-bond coupling between H– 3 and H–11 (or H–10 and H–4, respectively), but a three-bond coupling between H–10 and H–11 (or H–3 and H–4, respectively).

---

[1] A and **B** are pairs of enantiomers. In Figs. 5.14.1 and 5.14.2 only the R isomer is shown.

[2] The carbon designation combines pairs of equivalent atoms (cf. Fig. 5.14.1). The first number always corresponds to the right-hand part of the molecule (C–3, C–2, C–1, C–14, C–13, C–12 and C–11), the second to the left-hand part (C–10, C–9, C–8, C–7, C–6, C–5 and C–4).

The stereochemical assignment of the methylene protons can be accomplished using the NOE difference spectrum. In addition, this experiment answers the question as to which of the two configurations, **A** or **B**, is present in the molecule. Here the low-frequency H–2/H–9 signal ($\delta = 0.75$) is irradiated causing significant signal enhancement for the geminal H–2/H–9 and one of the H–13/H–6s as well as a weaker one for one of the H–14/H–7. This observation is consistent only with the syn configuration (**A**), since in the anti configuration (**B**) the relative intensities of the H–13/H–6 and H–14/H–7 signals should be reversed.

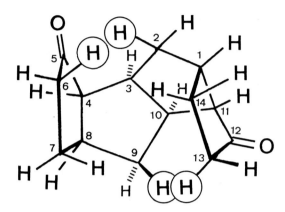

Fig. 5.14.2. Stereoprojection of the syn-configuration of **22**.

The irradiated proton can only be the endo-positioned[3] H–2n/H–9n, denoted by circles in Fig. 5.14.2. The signal splitting confirms the syn configuration; we can assume from the quartet-type appearance of the signal that H–2n/H–9n has three couplings of approximately the same magnitude. This is possible only if H–1/H–8 and H–3/H–10 are anti positioned. The atoms H–13/H–6 – also denoted by circles in Fig. 5.14.2 – and H–14/H–7, which are both affected by the NOE, adopt the endo position.

Finally, the two possible conformations have to be differentiated: A spatial proximity may exist between H–2n and H–6n (or H–9n and H–13n, respectively), that is, between hydrogens from different parts of the molecule. On the other hand, such proximity may also exist between H–2n and H–13n (or H–9n and H–6n, respectively), that is, between hydrogens from the same part. It can be deduced from the relative magnitudes of their NOEs that H–6n (or H– 13n, respectively) is clearly closer to the irradiated proton than H–7n (or H–14n, respectively). A Dreiding model shows that the only conformation without an exceptional molecular strain is the one in which the atoms from different parts of the molecule are close to each other. Therefore, it is reasonable to assume that the molecule favors a conformation in which the two terminal five-membered rings avoid each other. This, however, is only a speculation based on models, not a proof.

---

[3] Designations: "n" denotes *endo* and "x" is *exo*.

Such a conformation can also explain the unusual chemical shift of $H-2^n/H-9^n$. These atoms are situated above the σ plane of the carbonyl groups so that a significant shielding is expected, an argument supporting the above-mentioned conformation assumption.

Signal assignment for **22**: δ_H = 2.70 (H–1/–8), 0.75 (H–2^n/–9^n)[3], 1.80 (H–2^x/–9^x), 2.58 (H–3/–10), 2.42 (H–11/–4), 2.07 (H–13^n/–6^n), 2.05 (H–13^x/–6^x), 1.61 (H–14^n/–7^n), and 1.77 (H–14^x/–7^x); δ_C = 43.8 (C–1/–8), 34.4 (C–2/–9), 49.1 (C–3/–10), 52.5 (C–11/–4), 221.3 (C–12/–5), 37.5 (C–13/–6), and 23.7 (C–14/–7).

References

1. Harris RK (1983) Nuclear Magnetic Resonance Spectroscopy, a Physicochemical View. Pitman, London, p 193.

Exercise 15

It can be seen in the H,H COSY plot that the signals at δ = 7.51, 6.78, and 6.69, as well as those at δ = 7.11, 6.43, and 6.42, respectively, belong to separate spin systems of aromatic protons. In the first system the signal splitting patterns (one broad, one narrow, and one double doublet) prove a 1, 2, 4-substitution; in the second the spin system is of high order and there is partial signal overlap. Nevertheless, there is a great similarity to the first spin system, so we can assume that it is again the same 1, 2, 4-substitution. The two signals at δ = 6.69 and 6.42 are narrow, that is, the corresponding protons do not possess ortho-proton neighbors. In addition, they have relatively small chemical shifts, so it is reasonable to assume that each of them is positioned between two oxygen functionalities [1]. The ^{13}C NMR spectrum displays four signals of quarternary aromatic carbons with large chemical shifts (δ = 161.2, 160.6, 156.2, and 151.6), that is, they belong to oxygenated carbons.

Apparently, the molecule contains two fragments of the structure:

The aliphatic protons (δ = 5.50, 4.26, 3.61, and 3.57) form a consistent ABMX spin system. Although this is of high order, some conclusions can be drawn: The signal at δ = 5.50 corresponds to a proton in geminal position with respect to an oxygen and displays a splitting of 6 to 7 Hz, that is, this proton apparently has only one vicinal partner, namely, that with δ = 3.57. There is a second coupling partner (δ = 4.26), but the underlying coupling seems to be long-range. The nucleus at δ = 3.57 couples with two more protons (δ = 4.26 and 3.61), which are connected to each other by a large coupling as well, so we can assume that they are diastereotopic protons of one methylene group; in the ^{13}C NMR and DEPT spectra a corresponding CH_2 signal can be found. Therefore, the aliphatic fragment is

The three fragments have to be combined in such a way that a tetracyclic compound is created, since the molecule contains 11 double-bond equivalents, 9 of which are covered by the aromatic rings and the acetate group. Thus, a new fragment is obtained leading to a parent compound belonging to the pterocarpan series:

The question arises whether the two heterorings are cis or trans fused. The NOE experiment with irradiation at H−11a (Fig. 3.15.4 b) shows a significant signal enhancement for C−6a, proving the cis fusion.

Finally, the positions of the two substituents have to be identified. First, a connectivity between the aliphatic and the two aromatic parts has to be found. This can be done with the NOE difference spectra; Fig. 3.15.4b provides a proof for the spatial proximity of H−1 and H−11a. Irradiation of the methoxy protons (Fig. 3.15.4c) shows that this group is between H−2 and H−4 (δ = 6.43 and 6.42, respectively). The counterexperiment (Fig. 3.15.4d) is the irradiation of the acetate protons, proving that this group is situated between H−2 and H−4 (δ = 6.78 and 6.69, respectively). Here the signal enhancements are quite weak owing to the large proton-proton distances.

Compound **23** is the acetate of medicarpin [2-4], and its NMR spectra have been published [5].

The two H−6 hydrogen atoms can also be assigned. The proton with the smaller chemical shift (δ = 3.61), in contrast to the other, has two couplings of approximately the same magnitude, one with the geminal H−6 and the other with the antiperiplanar H−6a. Therefore, it is H−6β.

Signal assignment for **23**: δ_H = 7.51 (H−1), 6.78 (H−2), 6.69 (H−4), 4.26 (H−6α), 3.61 (H−6β), 3.57 (H−6a), 7.11 (H−7), 6.43 (H−8), 6.42 (H−10), 5.50 (H−11a), 3.75 (OCH$_3$), and 2.28 (OCOCH$_3$); δ_C^1 = 131.8 (C−1), 110.7 (C−2), 161.2, 160.6, 156.2, and 151.6 (C−3/4a/9/10a), 115.3 (C−4), 66.6 (C−6), 39.5 (C−6a), 118.8 and 117.8 (C−6b/11b), 124.8 (C−7), 106.5 (C−8), 96.9 (C−10), 78.0 (C−11a), 55.5 (OCH$_3$), 21.1 (acetyl-CH$_3$), and 169.2 (acetyl-C=O).

---

[1] The signal assignment is not completely unambiguous because of the lack of an H,C COSY spectrum. It is based on data from the literature [3,4] and comparisons with similar compounds.

References

1. Pretsch E, Clerc T, Seibl J, Simon W (1983) Tables of Spectral Data for Structure Determination of Organic Compounds. Springer, Berlin Heidelberg New York.
2. Pachler KGR, Underwood WGE (1967) *Tetrahedron* **23**: 1817.
3. McMurray TBH, Martin E, Donnelly DMX, Thompson JC (1972) *Phytochemistry* **11**: 3283.
4. Duddeck H, Yenesew A, Dagne E (1987) *Bull Chem Soc Ethiop* **1**: 36.
5. Al-Ani HAM, Dewick PW (1984) *J Chem Soc Perkin* I 2831.

Exercise 16

First, the 10 signals in the 2D INADEQUATE spectrum are labeled from left to right by the capital letters **A** through **K**[1]. **F** and **K** belong to the two methylene groups, which have different neighbors: C–5 is situated between one aliphatic and one olefinic methine fragment, whereas C–10 is between two aliphatic fragments.

An evaluation, as described for cyclooctanol in Sect. 2.7 (cf. Fig. 2.7.2), shows that atom **F** is connected to the carbons **G** and **H**, and atom **K** to **I** and one of the olefinic carbons **B, C,** or **D** (**B/C/D**) (a differentiation between **B, C,** and **D** is not yet possible owing to the large scale of Fig. 3.16.5). Thus, **F** is C–10 and **K** is C–5. **A** is a neighbor of **H**, and a second atom among **B/C/D** is a neighbor of **G**; local symmetry prevents an assignment of **H** and **G** to C–8 and C–9 at this stage. Each of **H** and **G** possesses one more neighbor, namely, **E** and **I**, respectively, and these two nuclei are connected to each other. The third ligand of **E** is the last of **B/C/D**, whereas the third ligand of **I** appears to be **K** (see above). Now, the $^{13}\text{C},^{13}\text{C}$ connectivity of **24** can be established to form

$$= \text{B/C/D} - \text{G} - \text{I} - \text{K} \diagdown \text{B/C/D} =$$
$$\diagdown \text{F} \quad | $$
$$= \text{A} - \text{H} - \text{E} - \text{B/C/D} =$$

There is no unequivocal proof for the connectivity within the double bonds. A cross peak indicating the existence of a bond between C–8 (**B/C/D**) and C–9 (**A**) is lacking because the experiment has been optimized for $^{13}\text{C},^{13}\text{C}$ one-bond coupling constants of about 35 Hz; for a coupling between two olefinic carbons, however, a value of 70 to 80 Hz is expected [1]. In addition, it cannot be seen whether the cross peak for the other double bond is missing too, since the corresponding chemical shifts are very similar and a possible cross peak would be very close to the diagonal.

The assignment of the three sp^2 carbons 3, 4, and 8 to the signals **B, C,** and **D** is not yet possible but can be achieved by inspecting the expanded section of the 2D INADEQUATE plot (Fig. 3.16.6). The two signals at the lefthand side are those diagonal signals visible in the lower lefthand corner of the full plot (Fig. 3.16.5). In the upper righthand corner of Fig. 3.16.6 three doublet-shaped cross peaks can be identified, and the digital resolution is good enough to show that they can be assigned to the three olefinic carbons signals of **B, C,** and **D**. The atom with the highest frequency – **B** with $\delta = 132.4$ – belongs to the neighbor of C–7; that is, it is C–8. Analogously, it can be determined that C–4 corresponds to the signal of atom **C** ($\delta = 132.2$) and C–3 to **D** ($\delta = 132.1$).

The H,C COSY provides the ^1H signal assignment, which can be checked by interpretation of the H,H COSY plot. It turns out that all cross peaks for couplings between vicinal protons in the methylene

---

[1]The chemical shifts in the 2D INADEQUATE spectrum differ a little from those in the 1D ^{13}C NMR spectrum. The reason is the highly concentrated solution used for the INADEQUATE experiment. In the text we prefer the values obtained from the 1D spectrum since these were taken from a sample with a normal, rather than excessively high, concentration. It can be taken for granted that the signal sequence in both experiments is the same.

and methine fragments can be identified. For instance, both H-10 and H-10' are coupled to H-1 and H-7. The two olefinic protons H-8 and H-9 can be differentiated as well, since the cross peaks for H-8/H-7 and H-9/H-1 are at significantly different levels.

An unambiguous stereochemical discrimination between the protons within each of the two pairs of methylene protons on the basis of the spectra depicted here is not possible but may be obtained from NOE difference experiments irradiating H-2 and H-6. Since, however, the proton chemical shift differences within each pair are very large, an assignment can be made on the basis of empirical knowledge and experience. H-5^n, as well as H-10, is situated within the shielding anisotropy region, but this does not hold to the same extent for H-5^x and H-10'[2]. Thus, it is expected that the latter have significantly larger chemical shifts in comparison with their respective geminal partners [2].

Signal assignment for **24**: δ_H = 2.66 (H-1), 3.12 (H-2), 5.48 (H-3 and H-4), 2.12 (H-5^x), 1.57 (H-5^n)[2], 2.58 (H-6), 2.73 (H-7), 5.92 (H-8), 5.94 (H-9), 1.16 (H-10), and 1.52 (H-10'); δ_C = 45.7 (C-1), 55.1 (C-2), 132.1 (C-3), 132.2 (C-4), 35.0 (C-5), 41.6 (C-6), 46.5 (C-7), 132.4 (C-8), 136.2 (C-9), 50.6 (C-10).

It is now possible to see how the ^{13}C signal assignments reported in the literature [3–5] differ from those derived here (Table 5.16.1). It can be seen that in all previous cases the signals of the olefinic carbons have been misassigned. Because all previous methods require interpretation, their information can be regarded only as more or less reliable hints; the 2D INADEQUATE spectrum provides an experimental proof.

Table 5.16.1. 13C signal assignments for 24.

Signal Source	Carbon/Signal No.									
	A	B	C	D	E	F	G	H	I	K
[3]	4	3	9	8	2	10	6	1	7	5
[4]	4	3	9	8	2	10	7	1	6	5
[5]	3	8	4	9	2	10	7	1	6	5
This work	9	8	4	3	2	10	7	1	6	5

References

1. Marshall JB (1983) Carbon-Carbon and Carbon-Proton NMR Couplings: Applications to Organic Stereochemistry and Conformational Analysis Methods in Stereochemical Analysis, Marchand AP (ed). Verlag Chemie International, Deerfield Beach, vol 2.
2. Harris RK (1983) Nuclear Magnetic Resonance Spectroscopy – A Physicochemical View. Pitman, London, p 193.
3. Johnson LF, Jankowski WC (1972) Carbon-13 NMR Spectra. Wiley, New York.
4. Sadtler Standard Carbon-13 NMR Spectra (1977). Sadtler Research Laboratories, Philadelphia.
5. Nakagawa K, Iwase S, Ishii Y, Hamanaka S, Okawa M (1977) *Bull Chem Soc Jpn* **50**: 2391.

---

[2] Designation: "n" denotes *endo*, "x" is *exo*.

Exercise 17

In the region of the sugar skeleton atoms (δ = 50 to 105) 11 ^{13}C signals can be found. Thus, the compound is a disaccharide consisting of one hexose (**H**) and one pentose (**P**). The chemical shifts of the two anomeric carbons (C–1) and the attached protons are δ_C = 96.6/δ_H = 4.91, and δ_C = 100.4/ δ_H = 4.55, respectively. Starting from each of these anomeric protons, all signals belonging to the same sugar unit can be identified in the H,H COSY spectrum, and the signal splitting due to ^{1}H,^{1}H couplings can be read (cf. Fig. 5.17.1). In addition, it is possible to collect the respective ^{13}C signals from the H,C COSY plot (Table 5.17.1).

Table 5.17.1. ^{1}H and ^{13}C Signals of **25**.

	Hexose (H)			Pentose (P)		
Fragment	δ_H	J	δ_C	δ_H	J	δ_C
1	4.91	J(1,2) = 3.6	96.6	4.55	J(1,2) = 6.6	100.4
2	4.82	J(2,3) = 10.3	70.8	5.16	J(2,3) = 9.0	68.9
3	5.43	J(3,4) = 9.3	70.4	5.03	J(3,4) = 3.5	69.9
4	5.05	J(4,5) = 10.0	68.7	5.21	d d d	67.4
5	3.88	J(5,6) = 4.7	68.4	3.98	J(4,5) = 3.8	62.7
		J(5,6′) = 2.0			J(5,6′) = 12.8	
5′				3.58	J(4,5′) = 2.0	
6	3.82	J(6,6′) = 11.5	66.0			
6′	3.66					

The remaining signals are OCH$_3$, δ_C = 55.4 and δ_H = 3.38; acetyl-methyl, δ_C = 20.9 (1CH$_3$), 20.8 (1CH$_3$), 20.7 (1CH$_3$), 20.6 (3CH$_3$) and δ_H = 2.09 (6H), 2.04 (3H), 2.02 (3H), 1.99 (3H).

The ^{1}H,^{1}H couplings J(1, 2), J(2, 3), J(3, 4), and J(4, 5) in the hexose part of the spectrum prove that H–1 is equatorial and H–2 through H–5 are axial in a pyranose ring[1]; this monosaccharide unit is an (α-glucoside. Analogously, it can be shown that the pentose is also a pyranose, in which atoms H–1 through H–4 are axial and H–4 is equatorial. This means that the second sugar moiety is an α-arabino-side. Here the H–5 proton with the larger chemical shift (δ = 3.98, H–5) is in the equatorial position, and the one with the smaller chemical shift (δ = 3.58, H–5′) is in the axial position.

The next questions are which of the two monosaccharides is methylated and which is attached to what position of the other via its anomeric carbon. This requires an experiment affording a connectivity of ^{1}H and/or ^{13}C nuclei from both monosaccharide subunits, for instance, a COLOC experiment (cf. Sect. 2.6) proving the existence of long-range ^{1}H, ^{13}C couplings. Often, however, only a limited amount of material is available, so the spectrometer time needed for a COLOC measurement would be an

---

[1] In general, coupling constants of 9 to 10 Hz do not appear in furanose rings.

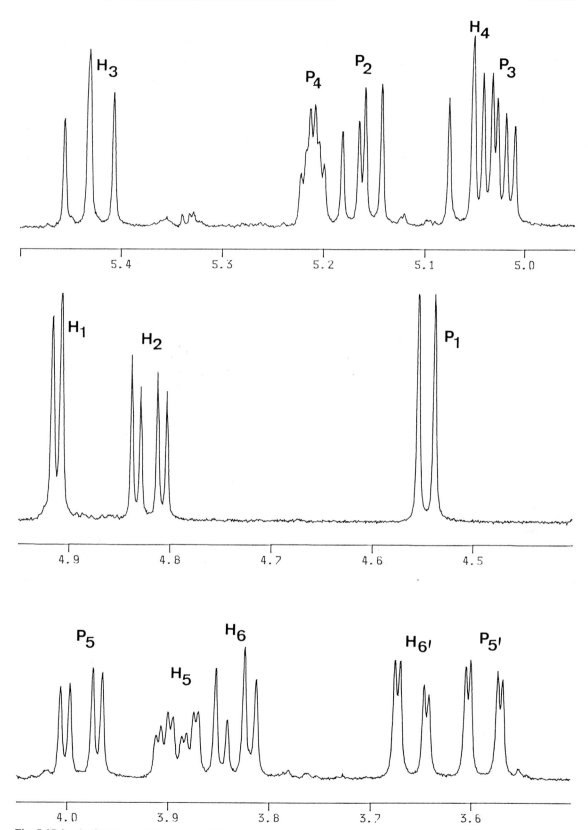

Fig. 5.17.1. Assignment of the sugar skeleton protons.

unsurmountable barrier. In the present case the proof of a single coupling (either H–1 of **H** with C–x of **P** or H–1 of **P** with C–x of **H**) would be sufficient answer to the above questions and can be provided by a 1D spectrum alone, namely, from a selective INEPT experiment (Fig. 3.17.2). The selective proton pulse is irradiated at $\delta = 4.55$, that is, at the H–1 signal of the arabinopyranoside moiety, and the parameters are optimized to a $^1H,^{13}C$ coupling of 5 Hz, thereby achieving a polarization transfer from that proton to C–6 of the glucopyranoside. Of course, the directly attached C–1 of the arabinose is affected too.

The disaccharide **25** is methyl-2,3,4-tri-O-acetyl-6-O-(2,3,4-tri-O-acetyl-α-D-arabinopyranosyl)-α-D-glucopyranoside.

References

For the NMR spectroscopy of carbohydrates:

1. Kotowycz G, Lemieux RU (1973) *Chem Rev* **73**: 669.
2. Wehrli FW, Nishida T (1979) *Progr Nat Prod Chem* **36**: 1.
3. Bradbury JH, Jenkins GA (1984) *Carbohydr Res* **126**: 125.
4. Davison BE (1985) *Carbohydr Chem* **16**: 224.

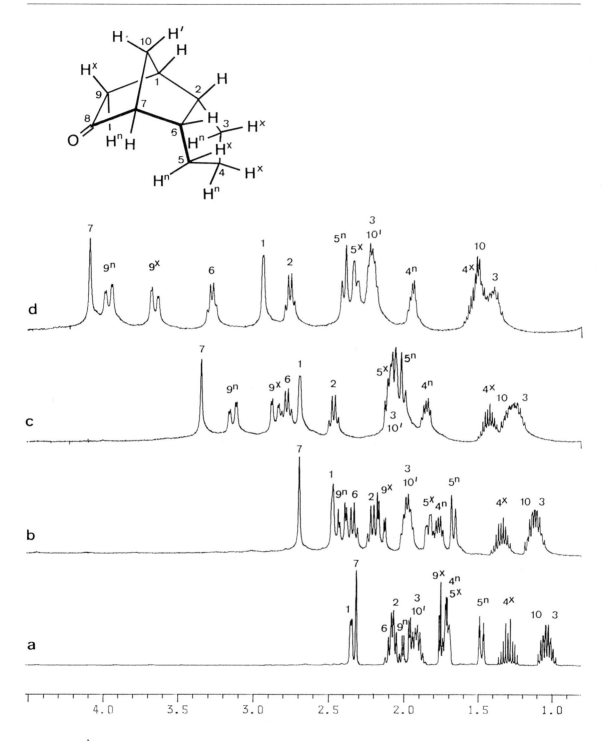

Fig. 5.20.2. ^{1}H NMR spectrum of **30** in the presence of the lanthanide shift reagent (LSR) Eu(fod-d_{27})$_3$: (a) Refe-rence spectrum without the addition of LSR. The molar ratio of LSR to **30** is (b) 0.05, (c) 0.10, (d) 0.15. All ^{1}H chemical shifts in (a) are larger by about 0.2 ppm as compared with Fig. 3.20.1. Such deviations are due to concentration differences. Designation: "n"denotes endo, "x"is exo.

Exercise 21

The ^{1}H and the ^{13}C NMR spectra clearly show that the complex is symmetrical; there are only three methyl and two methylene signals and one peak for a quarternary carbon. This is consistent with the fact that this dication has C_{2v} symmetry, where both ruthenium ions, the oxo bridge, and one N–CH$_3$ group lie on a mirror plane [1,2]. In the spectra the proton signals are labelled from left to right with lowercase letters (**a** through **i**) and those of the carbons by capital letters (**A** through **E**); the carbon signal with $\delta = 193.0$ is ignored because it obviously corresponds to the carbonyl group of the acetate bridges. It should be noted that, as compared with "normal" acetates ($\delta \approx 170$), this δ value is rather large owing to the bidentate character of the group. The corresponding acetate methyl signals are $\delta_C = 23.9$ and $\delta_H = 1.78$.

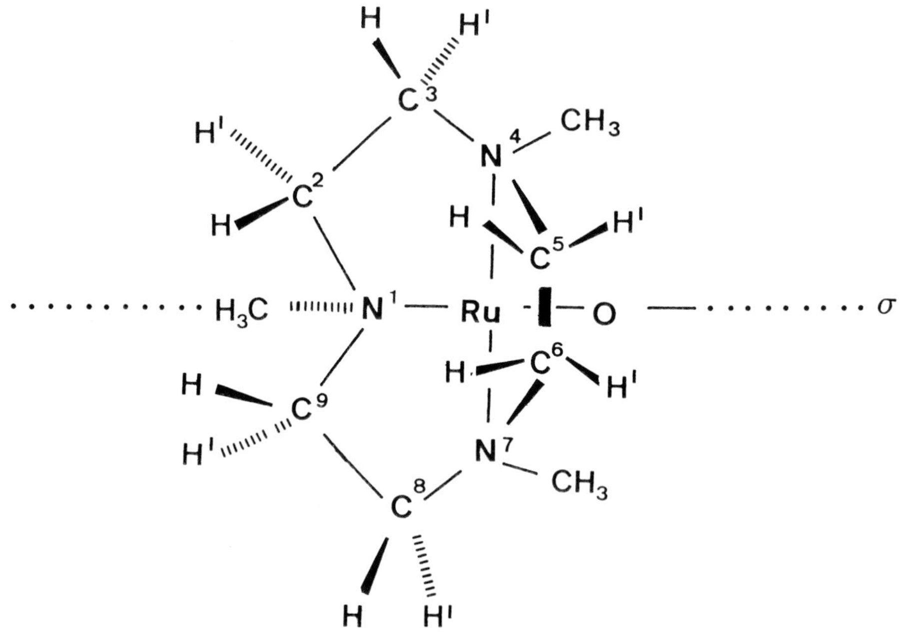

Fig. 5.21.1. Partial structure of **31**; the dotted line indicates the position of the mirror plane (sigma), which is perpendicular to the paper plane.

Differentiation of the two N-methyl groups is possible via the proton signal intensities. The signal at the lower frequency ($\delta = 1.35$) corresponds to six protons and that at $\delta = 3.77$ to only three. Thus the latter methyl group is the one lying in the mirror plane (N^1–CH$_3$).

In view of the molecular symmetry, one would expect three rather than two methylene carbon signals, and, indeed, the H,C COSY plot proves that the signal at $\delta = 63.8$ corresponds to two isochronous carbons. Moreover, this spectrum allows the attribution of signals to the individual methylene groups: **A-b-c**, **A-f-g**, and **B-a-e**. The first of these three fragments contains an ^{1}H spin system isolated from the others (cf. H,H COSY spectrum), and its appearance is characteristic of an AA'BB' subspectrum. Thus, it represents the ethylene bridge bisected by the mirror plane (C–5/C–6). The hydrogen nuclei of the remaining methylene groups constitute a four-spin system (AMNX) for the other two equivalent ethylene bridges (C–2/C–3 and C–8/C–9).

The signal assignments can be obtained from NOE difference experiments: Irradiation of signal **i** (N^4/N^7–CH$_3$, Fig. 3.21.4g) produces signal enhancements for **c** and **g**. Thus **c** belongs to H–5′/H–6′ and **g** to H–3′/H–8′. The respective geminal partners correspond to peaks **b** (H–5/H–6) and **f** (H–3/H–8). When the proton with $\delta = 4.53$ (**a**) is irradiated (Fig. 3.21.4b), NOEs can be observed for H–3/H–8 and, to a lesser extent, H–5/H–6; that is, the irradiated proton is H–2/H–9. The last remaining signal ($\delta = 3.66$) is that of H–2′/H–9′, and when this signal is irradiated, H–3′/H–8′ (**g**) responds significantly (Fig. 3.21.4e). Moreover, NOEs prove the spatial proximity of H–5/H–6 and H–3/H–8 (Fig. 3.21.4c), of N^1–CH$_3$ and H–2′/H–9′ (Fig. 3.21.4d), and of H–3′/H–8′ and N^4/N^7–CH$_3$ (Fig. 3.21.4f).

Two proton signals appear at an unusual chemical shift range. One of the methylene hydrogen atoms, namely, H–5′/H–6′, is more than 1.2 ppm lower in frequency than the nearest. The difference in the N-methyl δ values is even higher; whereas that of N^1–CH$_3$ ($\delta = 3.77$) is in the expected region, the methyl protons at N^4/N^7 are extraordinarily shielded ($\delta = 1.35$). It is striking that the shielded hydrogens are those directed toward the oxygen atom fixed in the central oxo bridge. An explanation may be the existence of strong anisotropic effects exerted by the free electron pairs of the oxygen, which come very close to the protons. This interpretation is confirmed by the fact that none of the respective carbons shows any significant diamagnetic signal shift, characteristic behavior when anisotropic effects are involved [3].

Signal assignment for **31**: $\delta_H = 4.53$ (H–2/H–9), 3.66 (H–2′/H–9′), 3.50 (H–3/H–8), 2.24 (H–3′/H–8′), 4.20 (H–5/H–6), 3.95 (H–5′/H–6′), 3.77 (N^1CH$_3$), 1.35 (N^4/N^7CH$_3$), 1.78 (CH$_3$COO), and ≈ 2.9 (H$_2$O); $\delta_C = 59.9$ (C–2/C–9), 63.8 (C–3/C–8 and C–5/C–6), 54.5 (N^1CH$_3$), 54.9 (N^4/N^7CH$_3$), 193.0 (CH$_3$CO2), and 23.9 (C$_3$COO$_2$).

References

1. Neubold P, Wieghardt K, Nuber B, Weiss J (1988) *Angew Chem* **100**: 990; *Angew Chem Int Ed Engl* **27**: 933.
2. Wieghardt K, personal communication.
3. Duddeck H (1986). In: Eliel EL, Wilen SH, Allinger NL (eds) Topics in Stereochemistry. Wiley, New York, vol 16, p 227.

Exercise 22

In this exercise possible ways of documenting the results of the evaluation of multipulse NMR spectra are demonstrated. For this reason the discussion is more detailed.

First, it is useful to mark all signals in the ^{1}H and ^{13}C NMR spectra (Fig. 5.22.1), from left to right, using capital letters for the carbons and lowercase letters for the hydrogens. The ^{1}H signal at δ 7.7 apparently belongs to a hydroxy group and is ignored in this context.

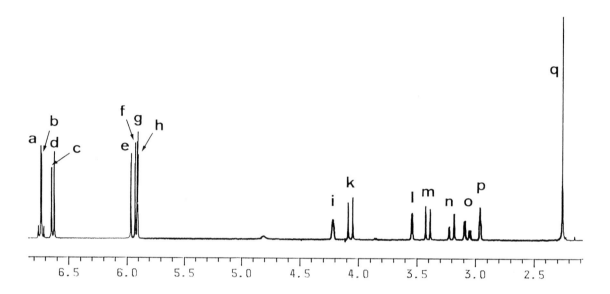

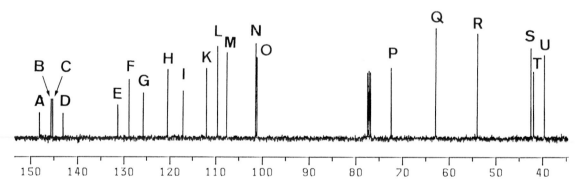

Fig. 5.22.1. Notation of the ^{1}H and ^{13}C signals of **32**.

The proton spectrum contains an AB system at the highest frequency (**a**: δ = 6.73; **b**: δ = 6.72) with a coupling of J_{HH} = 8.1 Hz, as well as two slightly broadened singlets at δ = 6.64 (**c**) and 6.62 (**d**). These four signals seem to belong to two different phenyl rings, one of them having the hydrogen in ortho and the other in para position. The signals at δ = 5.97 through 5.90 (**e** through **h**) belong to two methylene groups with **N** ($\delta = 101.1$) and **O** ($\delta = 100.9$); these chemical shifts suggest that they may correspond to olefinic $= CH_2$ groups. This would require two more sp^2 carbons, one for each $= CH_2$. Such atoms, however, are not available, since all four quarternary carbons resonating in that ^{13}C spectral region (**A** through **G** and **I**) are needed for the two aromatic rings. Thus, one has to assume that the two methylene groups are aliphatic with unusually high chemical shifts. The only plausible explanation for this finding is that each of the methylene fragments is attached to two oxygen atoms, forming dioxoles (formaldehyde acetals). Indeed, there are four aromatic quarternary carbons (**A** through **D**), the chemical shifts of which ($\delta = 147.9$, 145.4, 145.0, and 142.8, respectively) are characteristic of oxygenated carbons. As a result, the two aromatic fragments I and II can be established (Fig. 5.22.2):

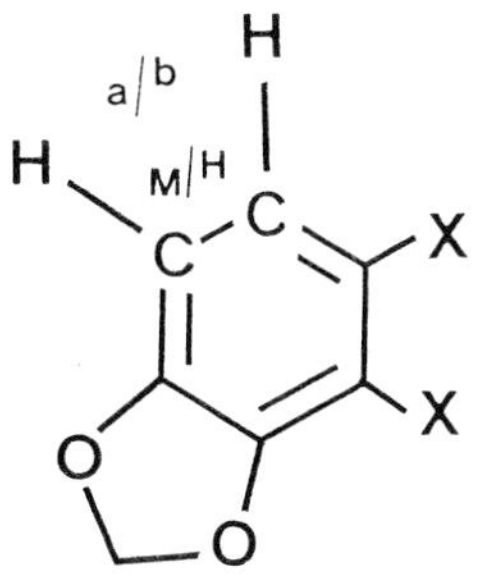

Fig. 5.22.2. Molecular fragments I and II.

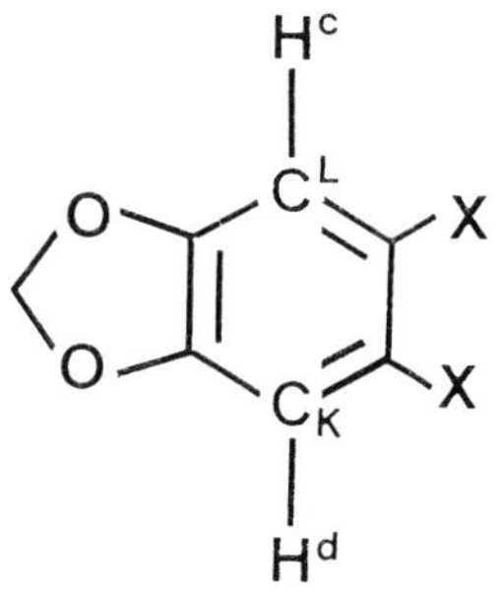

The ^1H NMR spectrum displays a singlet at $\delta = 2.25$ (**q**) with the intensity of three protons. The corresponding ^{13}C chemical shift (H,C COSY plot) is 42.3 (**S**). This is a methyl group, and the δ values indicate that it is attached to the only nitrogen atom in the molecule.

Further, seven aliphatic hydrogens (**i** through **p**) can be found, and according to the H,C COSY spectrum they correspond to three CH (**P**, **Q** and **T**) and two CH_2 fragments (**R** and **U**). The methylene group containing **R** is apparently isolated from the other aliphatic and aromatic fragments, since the attached hydrogens **k** ($\delta = 4.07$) and **m** ($\delta = 3.40$) form a two-spin-system without additional significant couplings; there are no other clear cross peaks in the H,H COSY plot.

This does not hold for the other methylene group with **U** ($\delta = 39.5$), **n** ($\delta = 3.18$), and **o** ($\delta = 3.07$). Both protons are coupled to **i** ($\delta = 4.21$). This chemical shift and the δ value of the directly bonded carbon **P** ($\delta = 72.2$) strongly indicate that **P** is substituted by an oxygen function, thus covering the last remaining oxygen atom present, according to the molecular formula. The IR spectrum proves the existence of a hydroxy group (compare the signal at $\delta \approx 7.7$), so a fragment with the structure $> CHOH$ emerges. Proton **i** is connected to both of the other methine hydrogens **l** ($\delta = 3.53$) and **p** ($\delta = 2.96$), which are interrelated by a clear cross peak in the H,H COSY plot. Atom **l** is bonded to the carbon **Q** with $\delta = 62.7$, a chemical shift indicating that **Q** is probably another ligand of the nitrogen atom. The last methine fragment, **T** through **p**, is situated between **P** and **Q**. The NOE difference experiment depicted in Fig. 3.22.4e, irradiation of proton **q** (N-methyl), confirms this arrangement of the carbon chain, since proton **l** is close to **q** but not to **p**. This argument, however, is premature at this stage because NOE experiments provide information about spatial proximity but do not prove connectivities in the molecules via bonds. Later in this discussion the NOE argument will turn out to be correct. Now we can establish another structural fragment – III (Fig. 5.22.3).

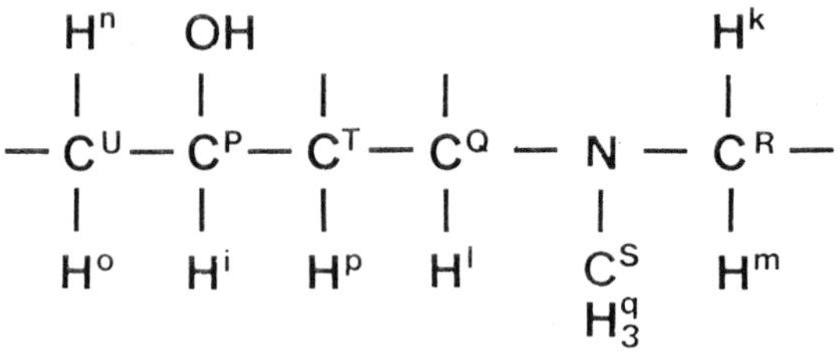

Fig. 5.22.3. Molecular fragment III.

The molecular formula indicates 12 double bond equivalents, 10 of which are already covered by the benzene residues (four each) and the dioxole rings. There are two more; that is, fragment III has to be connected to I and II through the formation of two more rings. This can create four isomers: The fragments may appear in the sequence I, III, II or II, III, I, and for the attachment of I to III there are two alternatives, one with the dioxole ring directed upward, and one downward. For the sake of space, only the formulas with the "downward-orientation" are depicted in Fig. 5.22.4.

A series of NOE difference experiments is able to show which of the four constitution formulas is correct: If the protons **n** and **o** are irradiated (Figs. 3.22.4b and 3.22.4c, respectively), there is a signal enhancement for the proton **c** in both cases. This proves the sequence I, III, II depicted in Fig. 5.22.4a. Correspondingly, proton **d** responds to the irradiation of the N-methyl hydrogens (Fig. 3.22.4e). The reduced scale of Fig. 3.22.4d, however, makes it difficult to see that the affected proton is indeed **d** and not **c**. On the other hand, the correct constitution is proved by the NOE difference spectrum displayed in Fig. 3.22.4d showing the spatial proximity of protons **p** and **a**.

We have now determined the constitution of the unknown compound **32**.

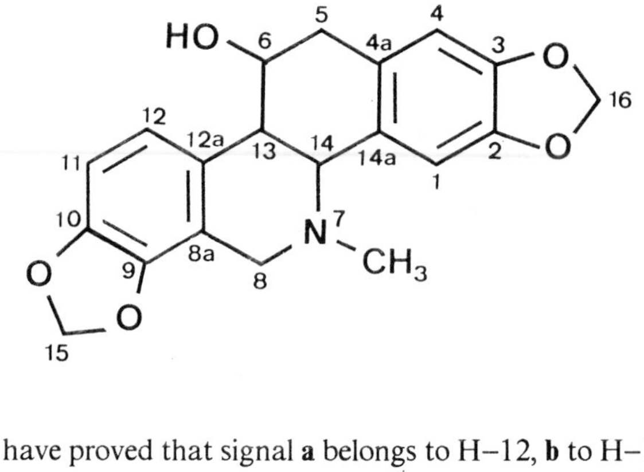

At the same time we have proved that signal **a** belongs to H–12, **b** to H–11, **M** to C–11, and **H** to C–12. The pairwise differentiation of the diastereotopic H–5 and H–8 is possible through inspection of the NOE difference spectra, as discussed later when the stereochemistry of **32** is being determined.

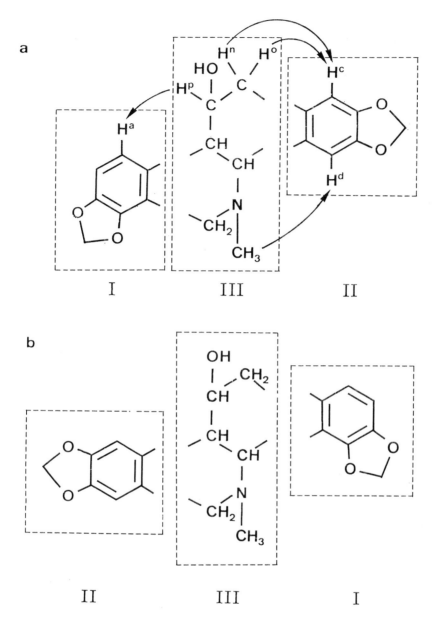

Fig. 5.22.4. Combining molecular fragments I, II, and III; the arrows indicate NOEs.

One way of clearly arranging the documentation for the ^{1}H,^{1}H connectivity of the aliphatic protons is presented in Table 5.22.1. For each cross peak a symbol "O" is registered in a matrix. In both the vertical and horizontal directions coupling partners for each nucleus can be readily recognized. Note that each signal is represented twice.

Table 5.22.1. Documentation of the Cross Peaks in the H,H COSY Spectrum of **32** (Aliphatic Part Only)

		6	8α	14	8β	5α	5β	13
		i	k	l	m	n	o	p
6	i			⊙		⊙	⊙	⊙
8α	k				⊙			
14	l	⊙						⊙
8β	m		⊙					
5α	n	⊙					⊙	
5β	o	⊙				⊙		
13	p	⊙		⊙				

Table 5.22.2. Documentation of the Cross Peaks in the H,C COSY (•) and in the COLOC Spectrum (⊙)

		3	10	2	9	12a	4a	14a	12	8a	1	4	11	15	16	6	14	8	Me	13	5
		A	B	C	D	E	F	G	H	I	K	L	M	N	O	P	Q	R	S	T	U
11	a		⊙		⊙	⊙							•								
12	b		⊙		⊙	⊙			•	⊙											
4	c			⊙			⊙					•									
1	d	⊙		⊙			⊙				•										
15	e				⊙									•							
16	f	⊙		⊙											•						
16	g	⊙		⊙											•						
15	h				⊙									•							
6	i						⊙									•					
8α	k				⊙	⊙				⊙							•				
14	l						⊙	⊙		⊙						⊙	•				
8β	m				⊙	⊙				⊙											
5α	n											⊙				⊙				⊙	•
5β	o						⊙	⊙				⊙				⊙				⊙	•
13	p				⊙				⊙	⊙										•	
Me	q																⊙	⊙	•		

The three ^{1}H signals **f, g, h** cannot be identified unambiguously in the given scale. An expansion of the COLOC spectrum suggests that **e** and **h** belong to one methylene group and **f** and **g** to another. A stereochemical discrimination of the proton within each methylene group is not possible with the available spectra. For structural elucidation, however, assignment is useless.

With the help of the COLOC plot, we can assign the signals of the quarternary carbons. For instance, the COLOC plot proves the existence of a long-range ^{1}H,^{13}C coupling between **A** and **d** (H–1), as well as between **A** and either **f** or **g**, signals that are very close together ($\delta = 5.93$ to 5.90; see Table 5.22.2). Thus **A** corresponds to C–3 and not to C–4 because the cross peak indicates a three-bond rather than a two-bond coupling (cf. Sect. 2.6). All other signals not yet assigned can be identified analogously. C–9 and C–10 can be differentiated by cross peaks connecting C–9 and the two H–8's. For the signal pair C–4a/C–14a differentiation is possible by finding the correlations between C–4a and H–1 or C–14a and H–4, respectively, both representing three-bond couplings. A discrimination between C–8a and C–12a is possible only via their chemical shifts. Increment calculations with regard to ortho and para effects in benzenes [1] lead us to expect C–8a to have a significantly smaller δ value than C–12a. The two –O–CH$_2$–O–protons the signal of which are at a distance from each other (**e** and **h**) display a connectivity with C–9 by COLOC cross peaks, that is, they are the two H–15 protons; thus, **f** and **g** are the two H–16's. For the assignment of ^{13}C signals in these fragments, an expansion of the H,C COSY plot is needed (Fig. 5.22.5). It can be seen that proton **e** is attached to the carbon N; that is, this signal corresponds to C–15.

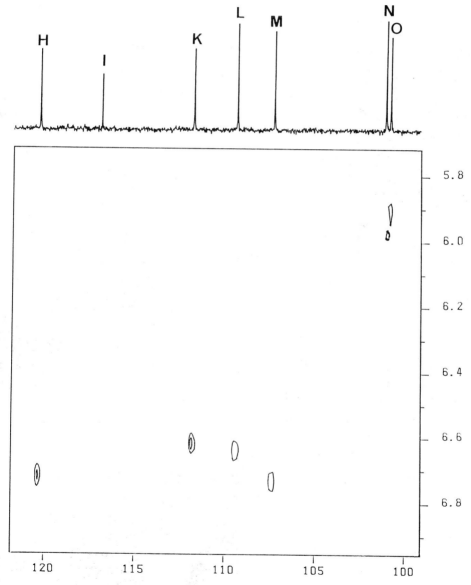

Fig. 5.22.5. Expanded section of the H,C COSY spectrum of **32**.

Finally, the stereochemistry of **32** has to be elucidated. The signal of H–6 is a multiplet, since this proton has four coupling partners (the two H–5 atoms, H–13, and H–14). Apparently, none of the coupling constants is large, thus eliminating the possibility of an antiperiplanar vicinal neighbor for H–6. This means that H–6 is in a quasi-equatorial and H–13 in a quasi-axial position. The NOE difference spectrum depicted in Fig. 3.22.4d proves the spatial proximity of H–13 and H–14; that is, both nuclei are on the same side of the molecule. Thus structures IV and V remain different, not only in their conformation, but also with regard to the configuration of C–6 (Fig. 5.22.6).

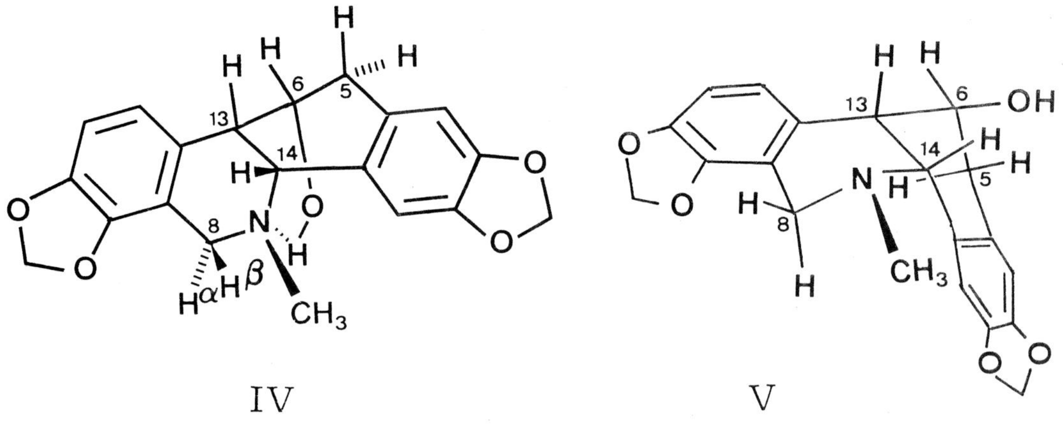

Fig. 5.22.6. Alternative structures of **32**.

It is easy to decide between IV and V since there are two independent arguments showing that **32** has the structure of formula IV: (a) As already mentioned, the molecule contains an intramolecular hydrogen bond [2], which explains the unusually large chemical shift of the hydroxyl proton ($\delta \approx 7.7$); such a hydrogen bond can exist only in IV. (b) There is a ^{1}H,^{1}H long-range coupling between H–6 and H–14, seen when the ^{1}H,^{1}H connectivity has been established. Such a coupling can give rise to a significant cross peak in the H,H COSY plot, but only if the two atoms involved are in a more or less coplanar W arrangement, as is the case only in IV.

Thus, compound **32** turns out to be chelidonin [3-5], an isoquinoline alkaloid belonging to the family of benzophenanthridines [2].

A pairwise differentiation of the two H–5 and two H–8 nuclei is possible too. Since there is no close spatial proximity between H–13 and one of the two H–5 atoms (NOE difference spectra in Figs. 3.22.4c and d), the methylene group containing C–5 must be rotated outward a little so that the quasi-axial H–5β on the β face of the molecule has a much smaller torsional angle with H–6 (approximately 30° to 40°, according to Dreiding models) than H–5α (approximately 90°). The corresponding coupling constants are expected to be a few hertz in the first case and very small in the second. This interpretation can be verified in the spectrum. The two H–8 protons can be discerned only by their chemical shifts. The quasi-equatorial H–8α is situated near the plane of the neighboring aromatic ring, the quasi-axial H–8β, however, is nearly orthogonal with respect to that plane. Owing to the ring current, it is expected that H–8α is significantly deshielded as compared with H–8β.

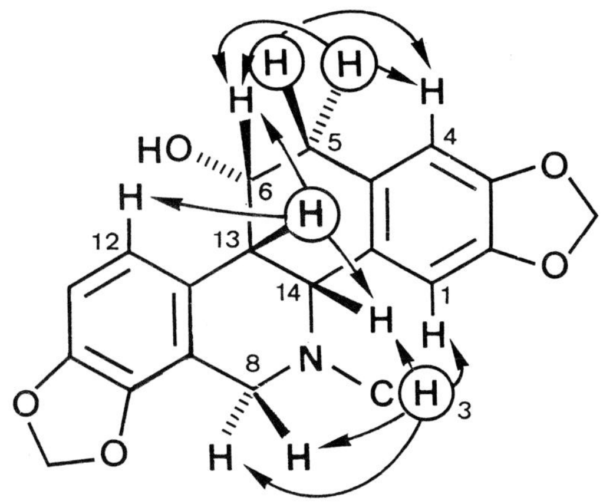

Fig. 5.22.7. Graphic representation of the NOEs in 32, irradiated nuclei are marked by circles.

Signal assignment for 32: δ_H = 6.62 (H-1), 6.64 (H-4), 3.18 (H-5α), 3.07 (H-5β), 4.21 (H-6), 4.07 (H-8α), 3.40 (H-8β), 6.73 (H-11), 6.72 (H-12), 2.96 (H-13), 3.53 (H-14), 5.97 and 5.90 (two H-15), 5.93 and 5.90 (two H-16), and 2.25 (NCH$_3$); δ_C = 111.8 (C-1), 145.0 (C-2), 147.9 (C-3), 109.4 (C-4), 128.6 (C-4a), 39.5 (C-5), 72.2 (C-6), 53.8 (C-8), 116.9 (C-8a), 142.8 (C-9), 145.4 (C-10), 107.3 (C-11), 120.3 (C-12), 131.1 (C-12a), 41.8 (C-13), 62.7 (C-14), 125.4 (C-14a), 101.1 (C-15), 100.9 (C-16), and 42.3 (N-CH$_3$).

References

1. Pretsch E, Clerc T, Seibl T, Simon W (1983) Tables of Spectral Data for Structure Determination of Organic Compounds. Springer, Berlin Heidelberg.
2. Bersch H-W (1958) *Arch Pharm* (Weinheim) **291**: 491.
3. Shamma M (1972) The Isoquinole Alkaloids. Academic Press, New York London, p 320.
4. For the conformation of chelidonin see: Naruto S, Arakawa S, Kaneko H (1968) *Tetrahedron Lett* 1705.
5. Snatzke G, Hrbek Jr. J, Hruban L, Horeau A, Santavy F (1970) *Tetrahedron* **26**: 5013.

Exercise 23

The following discussion of a solution pathway is again supplemented by documentation and is, therefore, especially detailed.

First, one should gain a general spectral overview: The ^{1}H NMR spectrum indicates the presence of several partially overlapping multiplets in the aliphatic region (δ = 1.5 to 2.1). However, all peaks are resolved in the ^{13}C NMR spectrum. Consequently, the HC COSY plot allows a straightforward determination of the proton resonances that are associated with the ^{13}C signals by one-bond coupling. An HH COSY experiment is performed to determine ^{1}H,^{1}H connectivity patterns. However, how the hydrocarbon fragments are linked together, cannot be established from either the homo- or the heteronuclear COSY experiments, since each end of these fragments is terminated by quarternary carbon or oxygen atoms (see below). This difficulty can be overcome by using the COLOC experiment, since it provides information about couplings of protons to the quarternary carbon atoms via more than one bond.

The overview is followed by the spectral evaluation procedure:

(a) Information from spectroscopic methods other than NMR includes the molecular formula, which is $C_{28}H_{34}O_7$. The molecule possesses 12 double bond equivalents (rings and/or multiple bonds), and according to the IR and UV spectra, it contains an α,β-unsaturated ketone and a furane ring (7).

(b) In examining the spectra for the number of signals, the ^{1}H and ^{13}C chemical shifts, and the proton signal shapes of **33**, we find five singlets at δ = 1.03, 1.04, 1.12, 1.19, and 1.22 for each three protons, multiplets between δ = 1.4 and 2.0 for six protons, one singlet at δ = 2.07 for three protons, one double doublet at δ = 2.12 for one proton, one double doublet at δ = 2.46 for one proton, one singlet at δ = 3.50 for one proton, one broad singlet at δ = 4.52 for one proton, one singlet at δ = 5.59 for one proton, one doublet at δ = 5.84 (J = 10.2 Hz) for one proton, one broad singlet at δ = 6.31 for one proton, one doublet at δ = 7.07 (J = 10.2 Hz) for one proton, and one doublet (or two singlets – splitting 1.2 Hz) at δ = 7.39 for two accidentally isochronous protons. The total number of signals accounts for all 34 hydrogen atoms encountered in the spectrum of **33**.

The ^{13}C NMR spectrum contains nine quarternary carbons: three at δ = 167.4, 169.9, and 204.0 (C=O), one at δ = 120.4 (olefinic), one at δ = 69.7 (attached to oxygen), and four at δ = 44.0, 42.6, 40.0, and 38.7 (aliphatic). Moreover, there are two CH fragments (δ = 46.0 and 39.5) for unsubstituted aliphatic carbon atoms, three CH fragments at δ = 56.8, 73.2, and 78.2 (attached to oxygen), five CH fragments in the olefinic/aromatic region (δ = 109.8, 125.9, 141.2, 143.1, and 157.0), three CH$_2$ fragments at δ = 14.9, 23.2 and 25.9, and six CH$_3$ fragments at δ = 17.7, 18.3, 19.7, 21.0, 21.2, and 27.1.

(c) The resonances are labeled by letters: capital letters for carbon and lowercase letters for hydrogens atoms. The sequence begins at the low frequency, that is, the right-hand-end of the spectra.

(d) The HH COSY signals are transferred from the contour plot (Figs. 3.23.2 and 3.23.5) to a correlation matrix (Fig. 5.23.3), which is a schematic image of the spectrum itself. (The weak cross peaks l–s, c–t, and a/b–u will later turn out to be artifacts, since significant couplings for these proton pairs cannot exist.) The heteronuclear experiments, HC COSY and COLOC, can be summarized in a single matrix (Fig. 5.23.4); filled dots represent HC COSY cross peaks, and open circles represent cross peaks from long-range ^{1}H,^{13}C couplings, as displayed in the COLOC spectrum.

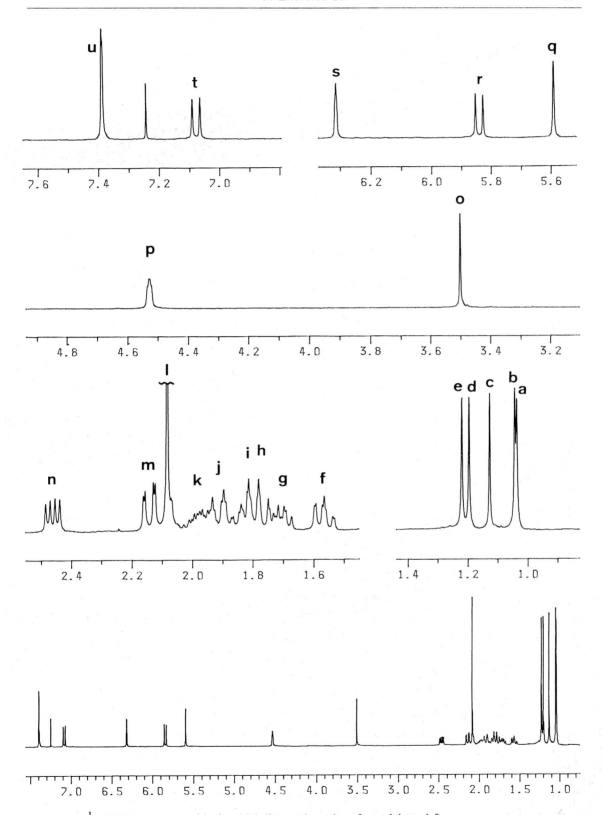

Fig. 5.23.1. ^{1}H NMR spectrum with signal labeling, a through u, from right to left.

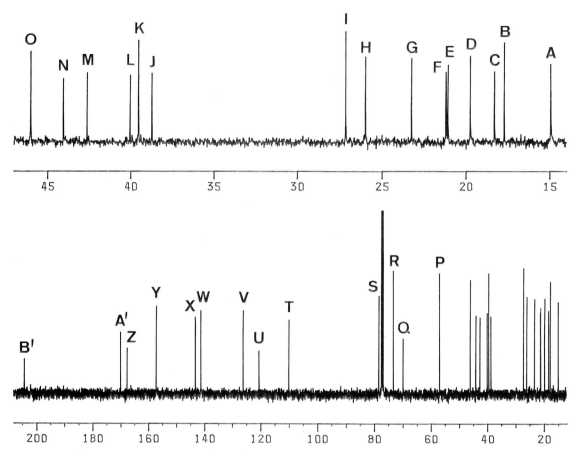

Fig. 5.23.2. ^{13}C NMR spectrum with signal labeling from right to left.

(e) The proton and carbon connectivities are established within the various $(CH_x)_y$ fragments that are separated from each other by quarternary carbon or oxygen atoms: Starting with carbon atom **A**, we can search for its filled dots in the HC matrix that depicts the directly attached protons; there are two along column **A** corresponding to protons **i** and **k**. In other words, we are dealing with a methylene group **A(i,k)**. Going down columns **i** and **k** in the HH matrix, we find that both have, among others, two coupling partners in common, namely, **f** and **n**. Proton **f** is attached to carbon **H** (see row **f** in the HC matrix) which is a methylene atom since it carries another hydrogen atom, namely, **g** (see column **H**), resulting in a second methylene group **H(f,g)**. Inspecting columns **f** and **g** in the HH matrix, we find no further coupling partners for these protons with significant cross peaks; that is, the fragment is terminated at this point. The cross peak **g–e** is very weak and may indicate a long-range coupling; such a peak, however, cannot be used for connectivity arguments at the present stage of the evaluation. Consequently, the fourth ligand of **H** must be a quarternary carbon (■) or an oxygen atom (Ω) (this symbol for oxygen is chosen in order to avoid confusion with carbon **O**). On the other hand, proton **n** belongs to the methine carbon **K**. The lack of additional vicinal proton partners for **n** indicates that **K** marks the other end of the fragment. The chemical shifts of **A**, **H**, and **K** suggest that none of them carries an oxygen atom; therefore, we can establish the molecular fragment **I** (Fig. 5.23.5).

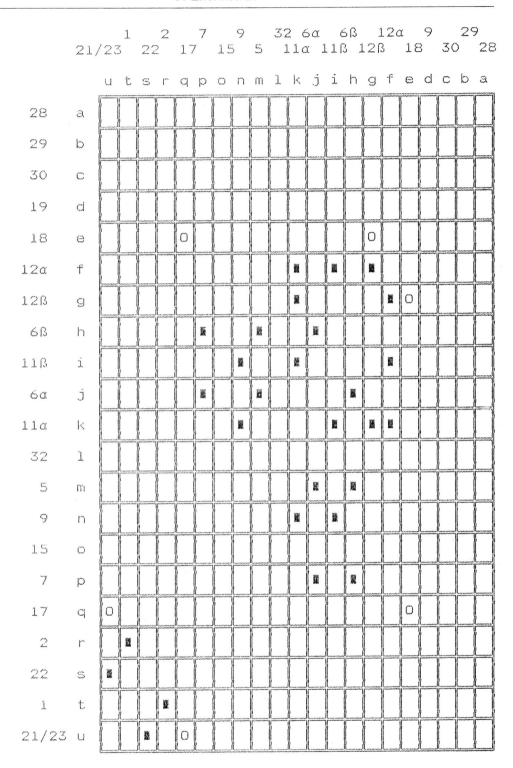

Fig. 5.23.3. ^{1}H,^{1}H connectivity matrix. The squares represent significant cross peaks due to geminal and vicinal couplings, and open circles represent very weak cross peaks due to vicinal couplings, with $J = 0$ to 1 Hz for long-range couplings.

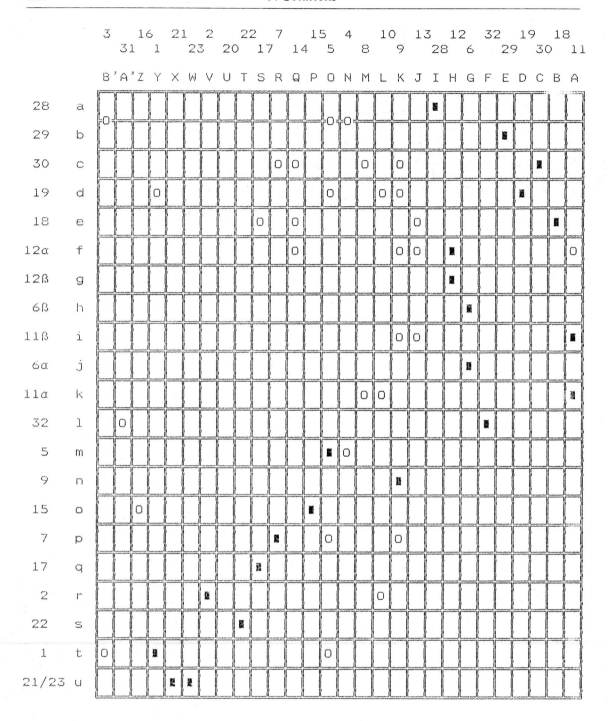

Fig. 5.23.4. ^{1}H,^{13}C connectivity matrix. The squares represent one-bond couplings (from the HC COSY spectrum), and circles represent multibond couplings (from the COLOC spectrum). The position of the three circles between the two rows for a and b indicate that it cannot be determined which of them is the coupling partner of the respective carbon atom.

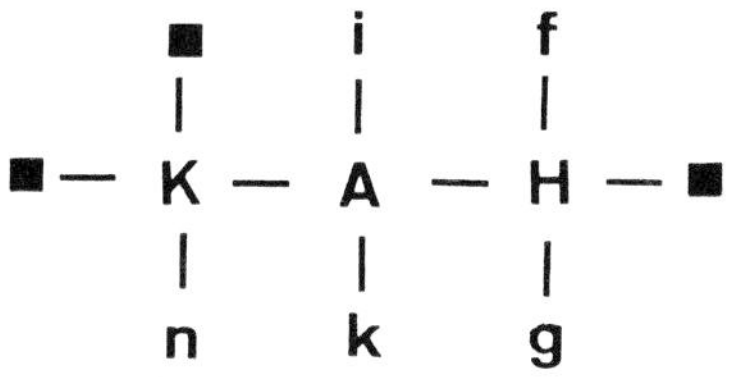

Fig. 5.23.5. Fragments established by proton and/or carbon connectivity arguments.

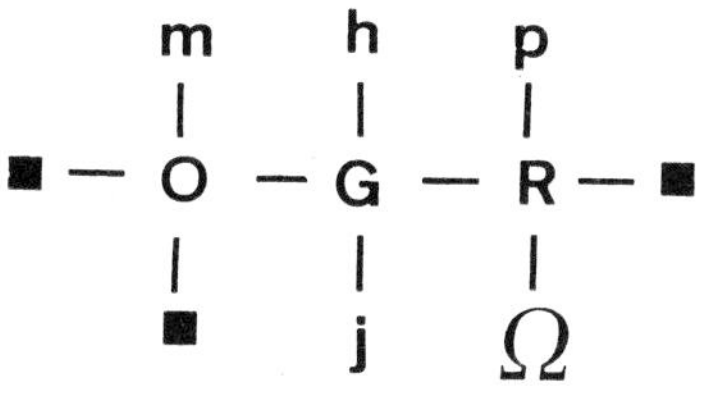

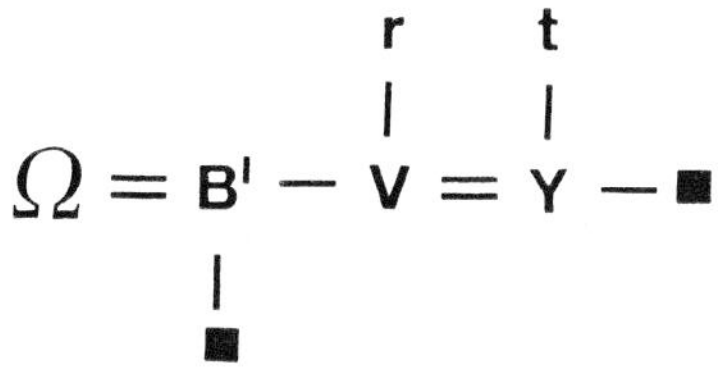

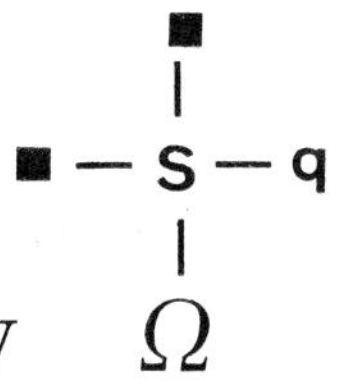

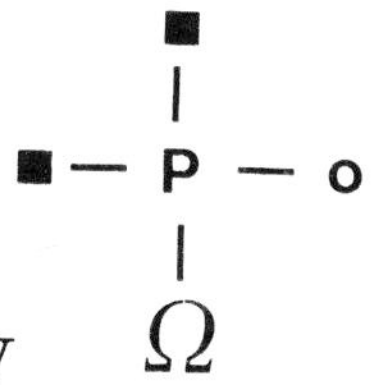

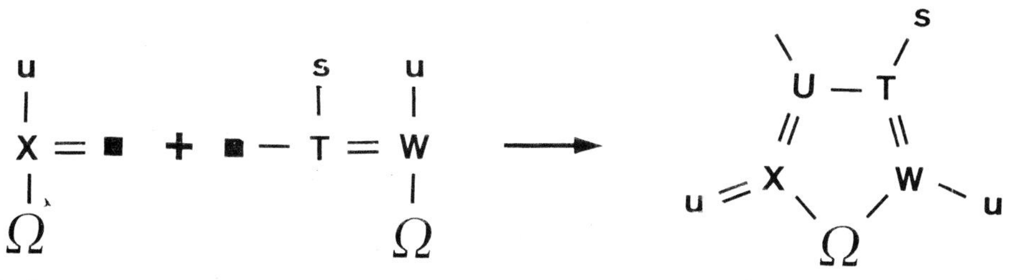

Repeating the same procedure leads to the formulation of five more fragments (**II** through **V** and **VIII**), all collected in Fig. 5.23.5. In fragment **III** carbon **V** apparently carries a C = O group since the ^{1}H and ^{13}C chemical shifts strongly suggest an α,β-unsaturated ketone, a group that has already been identified by other spectroscopic means (IR and UV). The chemical shifts indicate that at each carbon atom, **P, R, S, X,** and **W,** one ligand is an oxygen atom. The fragments **VI** and **VII** seem to be parts of the furane moiety, the presence of which has been suggested earlier (see above). The connecting quarternary olefinic carbon atom is easily identified: it is **U,** since this corresponds to the only quarternary carbon signal in the olefinic region ($\delta = 120.4$). Now the fragments **VI** and **VII** can be combined to form fragment **VIII** (Fig. 5.23.5);

(f) It is apparent that the two carbon signals at $\delta = 167.4$ and 169.9 are associated with two carboxylic acid functions. The chemical shift of the latter is characteristic for an acetate (**A'**). This is confirmed by two more signals that are typical for a methyl of an acetoxy group: proton **l** and carbon **F**.

(g) The chemical shift of one of the five aliphatic quarternary carbons (**Q**; $\delta = 69.7$) strongly suggests the existence of an attached oxygen atom.

(h) Now the jigsaw puzzle comprising seven double-bond equivalents is complete: Five are in fragments **III** and **VIII** and two in the above-mentioned carboxylic acid functions. With regard to the molecular formula we need five additional double bond equivalents; therefore, we have to connect the fragments by forming five more rings.

(i) The next step in the structural assignment is the interconnection of the fragments. In order to determine the connectivity of the molecular backbone, long-range ^{1}H,^{13}C coupling information extracted from the COLOC experiment is used:

As can be seen from HC matrix, the methyl carbon atom **D** with the three hydrogen atoms **d** fulfills all the necessary prerequisites of a "good" and diagnostically selective entry point into the core of the molecule. It exhibits the couplings of the **d** hydrogens with the quarternary carbon atom **L,** which, in turn, is directly connected to the three methine carbon atoms, **K, O,** and **Y,** that represent terminal carbon atoms in the **I, II,** and **III** fragments, respectively. The argument for these connectivities is as follows: Each of these carbon atoms has a coupling with the **d** hydrogen nuclei three bonds away ($^3J_{CH}$), represented by COLOC cross peaks. Similarly, **L** couples to **k** and **r**. Thus, we can construct fragment **IX** (Fig. 5.23.6).

Now, the first six-membered ring can be closed: the HC matrix displays couplings between the carbonyl atom **B'** and protons **a** and/or **b**, which are attached to the methyl carbon atoms **I** and **E,** respectively. Furthermore, the same methyl protons **a** and/or **b** are coupled to the quarternary carbon **N** and also to the methine carbon **O**. Thus, we find that the quarternary carbons attached to both **B'** and **O** must be identical, namely, **N,** carrying two geminal methyl groups (Ia_3 and Eb_3). This assignment is supported by an additional cross peak between **N** and **m**.

The second ring – again six membered – is identified by the following: Carbon **K** has a coupling with **p,** which is directly connected to **R,** proving the identity of the quarternary carbons attached to both **K** and **R** in fragment **IX**. Carbon **K** is also coupled to the methyl protons **c** at **C,** leading to the fourth angular methyl group. However, the atom between **C, K,** and **R,** which must be a quarternary carbon, is not yet identified. Among the quarternary carbons only two, namely, **M** and **Q,** show cou-

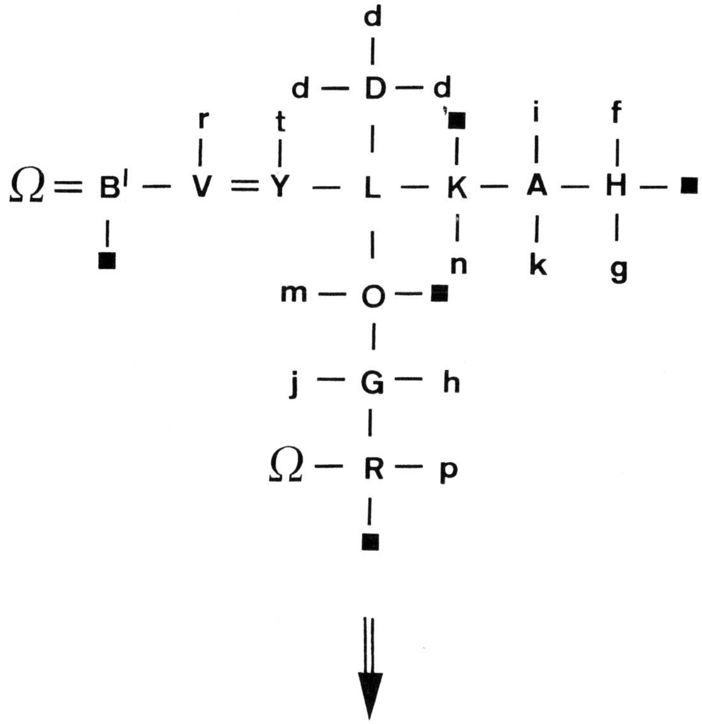

Fig. 5.23.6. Combining fragments I–XIII.

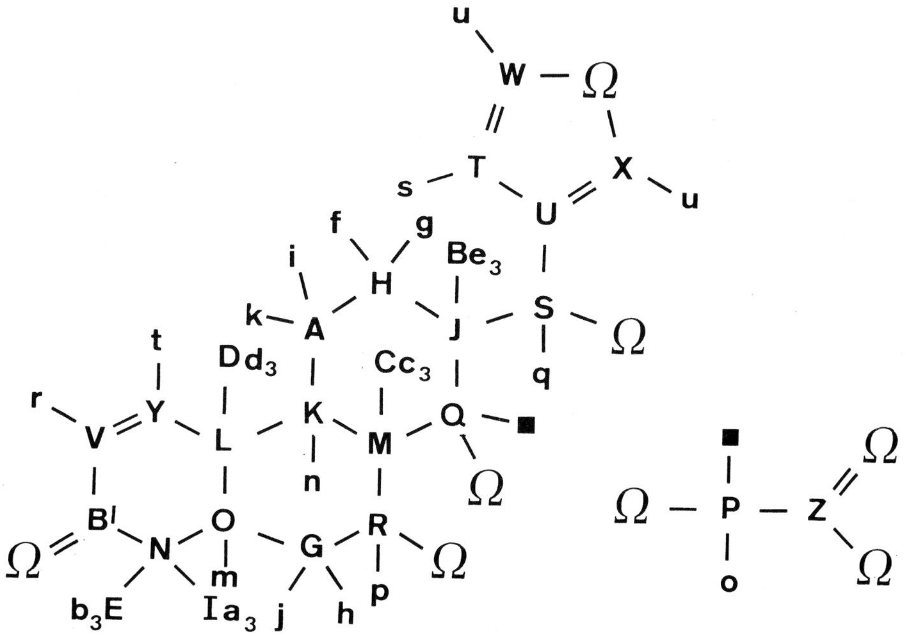

plings to the methyl protons **c**. Whereas **M** has an additional long-range coupling to proton **k**, **Q** couples to **e** and **f**. Proton **e**, however, is not present in the fragments discussed so far, and **f** is four bonds away, too remote for a reasonable COLOC peak. Thus, it is reasonable to conclude that the linking quarternary carbon atom is **M**.

Next is the formation of the third ring. Proton **f** couples with **J** and **Q**, both being quarternary carbons; the chemical shift of the latter suggests that it carries an oxygen atom. Because it is already ring ketone. There are nine additional ^{13}C signals, all corresponding to sp^3 hybridized carbons. This known that **Q** is not far from the methyl group Cc$_3$, **J** must be adjacent to **H**. **J** carries a methyl group consisting of **B** and three protons **e** (coupling between **J** and **e**). A further proof for the connectivity of **H** and **J**, is the fact that there is a four-bond coupling (apparently a **W** arrangement) between **e** and **g**. Further, we can connect **S**–**q** (fragment **IV**) to the carbon atom **J** because we can find a coupling between **e** and **S**. A weak HH COSY cross peak for a coupling between **q** and **u** suggests that the furane ring (fragment **VIII**) is connected to carbon **S**. This is nicely confirmed by the NOE difference spectra, which prove close spatial relationships between **q**, **s**, and **u**, as well as between **s** and **e**. Therefore, fragment **X** can be constructed (Fig. 5.23.6). The remaining atoms can be combined with the ester fragment **XI**, since a COLOC peak for **Z**–**o** can be identified. Finally, an acetyl residue has to be attached.

At this stage it cannot be decided unambiguously what the constitution of the molecule is. There are two possible alternatives (**XII** or **XIII**, Fig. 5.23.7), taking into account that two more rings have to be formed. The decision can only be made on the basis of the following NOE difference evidence. There is a close spatial relationship between the protons **p** and **o**, and this is only compatible with alternative **XII**, representing the skeleton of a limonoid derivative.

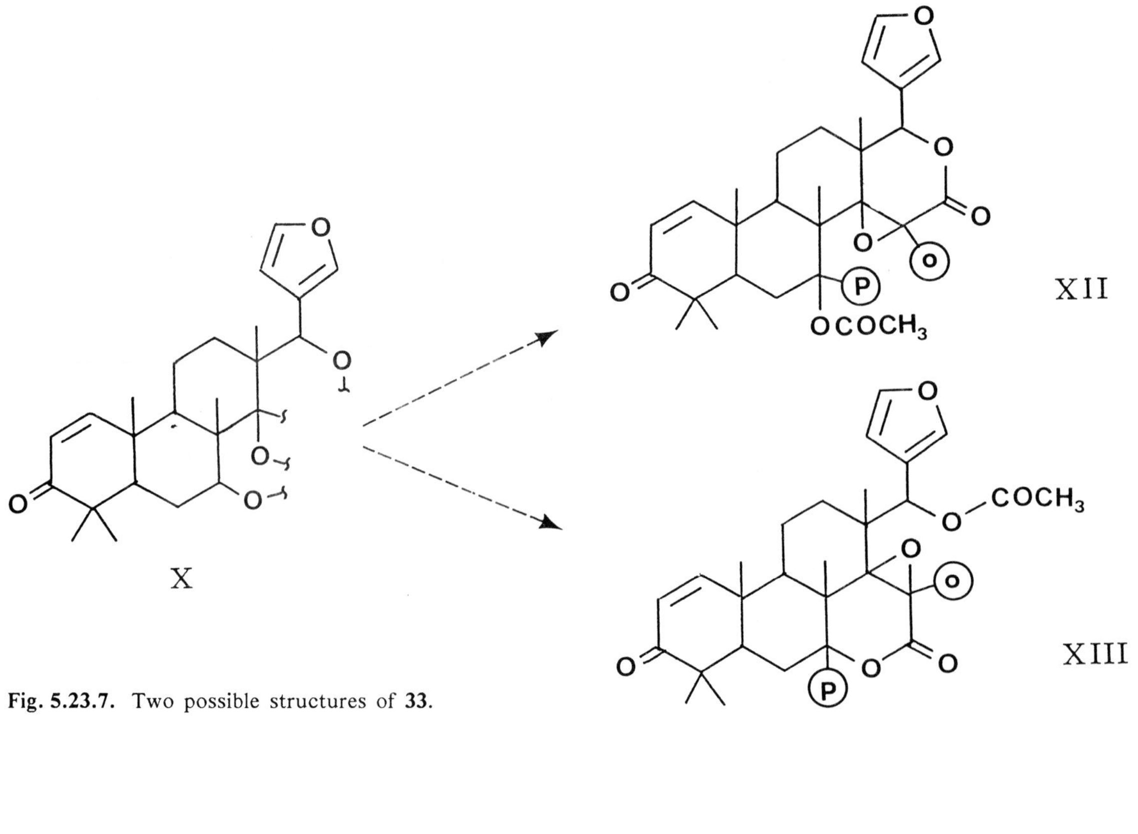

Fig. 5.23.7. Two possible structures of **33**.

(j) With the aid of the NOE difference spectra, as represented by the the NOE matrix (Fig. 5.23.8), the stereochemistry of the compound is established as having constitution **XII**.

Owing to the small distances between the methyl resonances in the ^{1}H NMR spectrum, the protons may be irradiated selectively (except for **a** and **b**), but the NOE responses cannot be detected for certain, and therefore the NOEs between methyl groups cannot be used for configuration determinations.

Taking into account the well-known absolute configurations of related limonoidal compounds, it is reasonable to assume that proton **m** is in the α position. The irradiation of this nucleus proves that protons **n** and **a** are on the α side as well; the absence of NOEs at **b**, **c**, and **d** strongly suggests that the methyl groups are in the β position. Thus, the trans configuration at the ring junction **L–O** is proved. Moreover, this experiment allows a distinction between the two geminal methyl groups at carbon **N**. No signal enhancement is observed when proton **n** is irradiated, so the methyl group Cc_3 is in β position proving the trans configuration of the second ring junction **K–M**. There is, however, a significant NOE at the methyl protons **e** indicating that this group (Be_3) is on the α side of the molecule. This is only possible if the third carbocyclic ring adopts a twist/boat conformation.

The position of the acetate group at carbon R is α, since proton **p** is not close to the methyl group Oc_3, and to both protons at carbon **G** (**j** and **h**).

The α position of the furane ring is proved by the fact that irradiation at the signal of proton **s** in that ring induces an NOE response on the methyl protons **e**, and vice versa, irradiation of **e**, NOE at **s** indicating their stereochemical proximity.

At this stage, it is possible to assign the geminal protons within each of the three methylene groups.

G(j, h): When **p** is irradiated, the signal shapes of both protons **h** and **j** can be observed without any overlap by other signals. The triplet form of **h** and the doublet form of **j** prove that **h** has two coupling partners with large **J** values (10 to 15 Hz), namely, **j** and **m**, whereas **j** has only one (**h**). This demonstrates that **m** and **h** are antiperiplanar with respect to each other, that is, that **h** is in β and, consequently, **j** is in α position.

A(i, k): When the protons c or d are irradiated, only proton **i** gives NOE responses; that is, **i** is on the β side.

H(f, g): When proton **f** is irradiated, significant NOEs are observed for **e**, **k**, **s**, and **u**, all of which are on the α side of the molecule.

The last configurational assignment, the orientation of the oxygen atom in the oxirane ring, is not easy. Molecular models show that the relative distances of protons **o** to **p** and **c** are more or less independent of the oxygen position. In the α-oxygen case the α-positioned proton **f** should be close to both protons **i** and **k** of the neighboring methylene group. The NOE experiment with irradiation at the signal for **f**, however, gives a signal intensity enhancement only for proton **k**, indicating that the other, **i**, is relatively far away from **f**. This observation exactly meets the expectations derived from the molecular model if the β orientation of the oxirane ring is assumed.

In conclusion, the relative configuration of the molecule is as depicted in Fig. 5.23.9, displaying the structure of a compound named gedunin, which has antimalarial activity [1]. It should be noted that the absolute configuration cannot be determined by the NMR methods described here because no chiral reference is available.

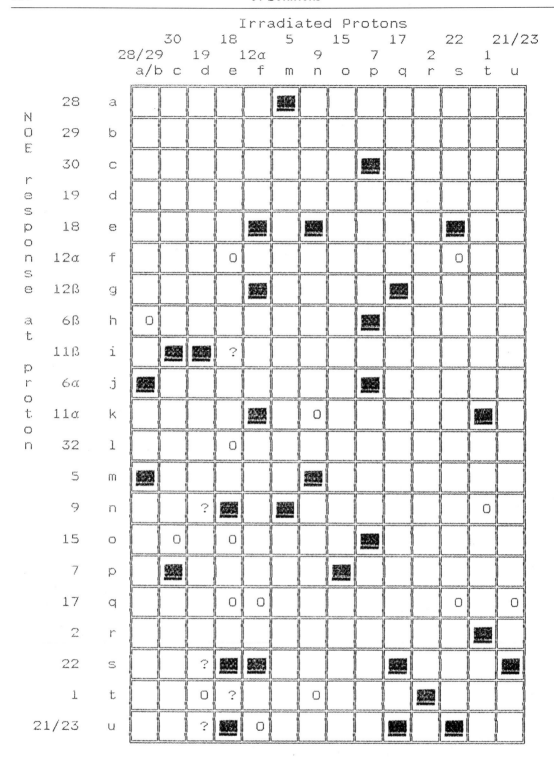

Fig. 5.23.8. NOE graph (arrows are directed from the irradiated to the affected protons: For the sake of clarity only a selection of diagnostically valuable NOEs is depicted): NOE matrix: Squares represent significant NOE difference signals, open circles weak ones; ? denotes signal responses that are apparently caused by an irradiation spillover from protons with resonances close to the irradiation position.

Fig. 5.23.8. (continued)

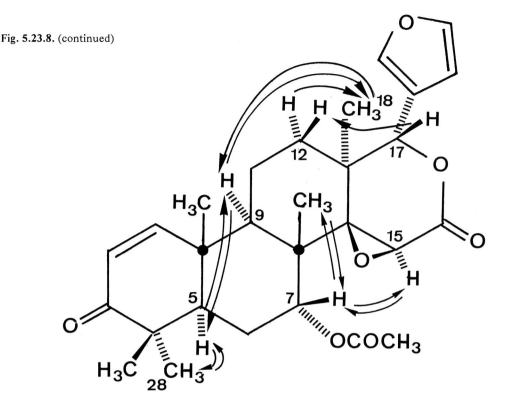

The ¹H and ¹³C chemical shifts of gedunin are collected in Table 5.23.1, the assignments of the atom letters to the numbers, according to Fig. 5.23.9, are also given and can be found in the peak matrices (Figs. 5.23.3, 5.23.4, and 5.23.8) as well. Our data confirm signal assignments published earlier for similar compounds [2].

This example impressively demonstrates that, instead of depending on empirical hints, a complete ¹H and ¹³C signal assignment of complicated structures can be reliably based on spectroscopic proofs. Previously, that is, using only one-dimensional NMR information, the spectroscopist had to rely largely on assumptions and speculations (the validity of which could only be estimated) derived from empirical knowledge and experience with related compounds.

Table 5.23.1. Signal assignments for **33**:

References

1. Khalid SA, Farouk A, Geary TG, Jensen JB (1986) *J Ethnopharmacol* **15**: 201.
2. Taylor DAH (1977) *J Chem Res (S)* **2**, (M) 0144; Kraus W, Cramer R, Sawitzki G (1981) *Phytochemistry* **20**: 117.

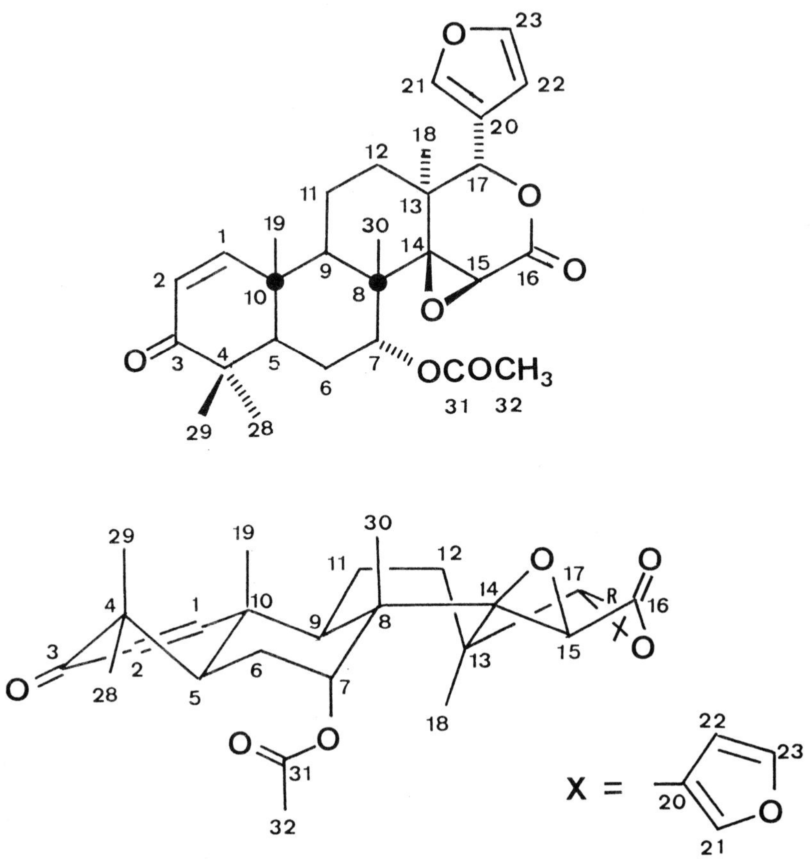

Fig. 5.23.9. Structure of gedunin (**33**).

Table 5.23.1. [13]C and [1]H Chemical Shifts of Gedunin, in CDCl3

The atom numbers refer to Fig. 5.23.9.

	[13]C				[1]H	
1	Y	157.0	1	t	7.07	(d, $J = 10.2$)
2	V	125.9	2	r	5.84	(d, $J = 10.2$)
3	B′	204.0				
4	N	44.0				
5	O	46.0	5	m	2.12	(dd, $J = 13.2, 2.3$)
6	G	23.2	6α	j	1.92	(d, $J \approx 12$)
			6β	h	1.79	(t, $J \approx 12$)
7	R	73.2	7	p	4.52	(br s)
8	M	42.6				
9	K	39.5	9	p	2.46	(dd, $J = 12.7, 6.2$)
10	L	40.0				
11	A	14.9	11α	k	2.00	(m)
			11β	i	1.81	(m)
12	H	25.9	12α	f	1.56	(dd, $J = 11$-12)
			12β	g	1.70	(m)
13	J	38.7				
14	Q	69.7				
15	P	56.8	15	o	3.50	(s)
16	Z	167.4				
17	S	78.2	17	q	5.59	(s)
18	B	17.7	18	e	1.22	(s)
19	D	19.7	19	d	1.19	(s)
20	U	120.4				
21	X	143.1	21	u	7.39	(d, $J = 1.3$)
22	T	109.8	22	s	6.31	(dd, $J \approx 1.3$)
23	W	141.2	23	u	7.39	(d, $J = 1.3$)
28	I	27.1	28	a	1.03	(s)
29	E	21.2	29	b	1.04	(s)
30	C	18.3	30	c	1.12	(s)
31	A′	169.9				
32	F	21.0	32	l	2.07	(s)

Compound Index

Exercise 1

Exercise 2

Exercise 3

Exercise 4

Exercise 5

HOCH2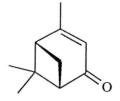

Exercise 6

Exercise 7

Exercise 8

Exercise 9

Exercise 10

Exercise 11

Exercise 12

Exercise 13

AcO OAc

AcO

O

O

O₂N   NO₂

Exercise 14

O

H H   H

H   H H

H

O

Exercise 15

O

O

O

H

CH₃

H

O

O — CH₃

Exercise 16

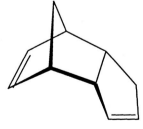

Exercise 17

Exercise 18

Exercise 19

Exercise 20

Exercise 21

$$\left. \right]^{2+} (PF_6^-)_2$$

Exercise 22

Exercise 23